User Design

User Design

Alison A. Carr-Chellman, PhD
Pennsylvania State University

LAWRENCE ERLBAUM ASSOCIATES, PUBLISHERS
2007   Mahwah, New Jersey   London

Copyright © 2007 by Lawrence Erlbaum Associates, Inc.
All rights reserved. No part of this book may be reproduced in any form, by photostat, microform, retrieval system, or any other means, without prior written permission of the publisher.

Lawrence Erlbaum Associates, Inc., Publishers
10 Industrial Avenue
Mahwah, New Jersey 07430

Cover design by Tomai Maridou

CIP information for this book can be obtained by contacting the Library of Congress

ISBN 0-8058-5504-1 (cloth : alk. paper)
ISBN 0-8058-5505-X (pbk. : alk. paper)

Books published by Lawrence Erlbaum Associates are printed on acid-free paper, and their bindings are chosen for strength and durability

Printed in the United States of America
10 9 8 7 6 5 4 3 2 1

This book is dedicated to the front line users in my life; Asher, Jules, and Aila. May you always be empowered to create your own futures.

Contents

Preface ix
Acknowledgments xi

1 User-Design: The Basics 1
2 History and Research 14
3 Tools for User-Design 25
4 Facilitating User-Design 45
5 Conflict 57
6 Leadership 77
7 User-Design and Systemic Change 85
8 User-Design and Performance Technology 99
9 Linking User-Design to Traditional Instructional Systems Design Models 106

References 125
Author Index 137
Subject Index 140

Preface

User design is a relatively new phenomenon that was first introduced into the consciousness of the Instructional Systems/Educational Technology field by Bela Banathy (1991). It is a construct that is closest to my heart because I believe in it. I believe in empowering users, people nearest to the ground, to the front lines. Learners have a right to design their own systems of human learning, and they have a responsibility to do so. Bela used to tell us that it is immoral to design for another. I believed him then, and I do today.

This book is a first foray into the world of user-design for instructional designers. It is organized according to the major issues associated with user-design. The first chapter introduces the basic concepts associated with user-design and explores the differences between user-design and stakeholder involvement and user-*centered* design. It examines the potential impacts of this movement on the field of instructional systems design ID.

The second chapter examines the history of the field of user-design and the empirical research associated with user-design. Understanding the theoretical foundations roots us in the knowledge of what has come before and what has been written in related fields such as human–computer interface design. The third chapter identifies five tools to be considered for a user-design approach including ethnography, cooperative design, action research-based user-design, design-based research, and scenario planning. In this chapter, guidance for when to select each tool and how to employ them is illustrated through fictional scenarios.

The fourth chapter outlines ways of facilitating user-design for anyone wishing to employ a user-design approach. Specifics for how to implement user-design in corporations and schools are drawn from two real-world case studies of the application of user-design in a home nursing agency and a suburban public school district. Implications from these cases are drawn for facilitation guidance. Chapter 5 is concerned with the care and feeding of leaders. Often user-design is a grassroots effort, but in order for it to reach maximum effectiveness, it must have the support, at some level, of the leadership in any organization. Therefore, learning how best to work with leaders and negotiate the pros and cons of user-design in an organizational context is crucial for the user-design facilitator.

Chapter 6 is focused on the conflicts that will inevitably arise during the course of any sincere user-design engagement. The reality is that users will bring different goals, agendas, and personal needs to the process; and so, it is very important for facilitators to see how to release the *energies* of conflict rather than learning how to handle, or avoid, or calm conflict. Chapter 7 presents a foundation in systems thinking and challenges you to become a systems thinker. Embedding user-design inside of a context of systemic change is essential to the appropriate and powerful outcomes of user empowerment.

Chapter 8 presents a special case of systems thinking and systemic change in the form of Performance Technology (PT). PT can be understood as the equivalent of systemic thinking in corporations, but the ways that a user-design facilitator might handle systemic change in a corporation has some specific differences from nonprofits, higher education, and public schools where systemic change is the more typical language. The final chapter integrates the ID process with the user-design process and lays out a typical day in the life of a traditional ID practitioner and contrasts that with the typical day in the life of a user-design ID practitioner. This helps novice user-design practitioners align and integrate user-design with ID.

Thought questions are posed at the end of each chapter as a way to help you synthesize the most important points and to apply them to your own context. To the extent possible, it will be helpful for you when approaching these questions to consider a single setting—typically either a school, corporation, or other organization such as a university, nonprofit, museum, etc. As you work your way through each chapter's questions, you'll find that keeping this single instance in mind will help you to build a clearer understanding of how user-design works within your context of interest better than if you jump around thinking about lots of different contexts, or thinking of some theoretical contexts (unless asked to do so).

This book assumes that you have a basic working knowledge of instructional design. While the fundamental ideas of ID will be covered again throughout the book, and get particular attention in the final chapter, the application of user-design to instructional design will be significantly aided if you are already familiar with the essence of the instructional design processes and theoretical underpinnings. In many ways, this text is meant to be used as a companion text within the learning of instructional design. There is nothing in this text which suggests that you should eschew the ID process in totem. Indeed, I agree wholeheartedly with Hannum (2005) when he writes, "There is nothing inherent in the ID model that requires us to discard it at this point" (p. 20). The purpose of this text is to broaden your perspective; to give you a new set of ideas and strategies that will dramatically impact the adoption rates of good designs. It presents a very different, and exciting, alternative to traditional instructional design. And it can also be seen as an approach that can be used in modification within the existing instructional design models despite its very different fundamental value structures.

I hope you enjoy your first steps into the exciting and challenging world of user-design. Whether you come to believe, as I have, that it represents the most moral approach to design, or if you merely find some of the tools useful within a traditional ID practice, I am confident that your design skills will improve dramatically as a result of learning more about user-design.

ACKNOWLEDGMENTS

Any authored book is the result of many people's efforts and help, and this one surely is no exception. First and foremost, I acknowledge the immense role that Dr Charles M. Reigeluth has played in this book and in my developing career path that has led to user design as a primary construct in my work. His guidance and gentle manner have been an inspiration to me, and I am deeply grateful for all of his intellectual direction. It was Dr Reigeluth who led me to Dr Bela Banathy both as an author and as a person. I acknowledge Bela's work as foundational to all that is created in this book. Bela's spirit will be with us always, and his moral ideals should be guideposts for those of us throughout the field of educational systems design and instructional systems.

I also thank those who have helped me to find the time in a busy life to complete this work. First on that list is most surely my husband, Davin, who, although managing his own academic career, still found time to help with our three little ones and to switch from changing diapers to deep intellectual discourse about the contents of this book in the blink of an eye. His discussions throughout the writing of this book have been instrumental in its basic structure and approach. I am also very deeply grateful to Barbara and Alva Chellman and Jack and Susan Keck for making this book possible with their daily support and love for our children. Without the help of four generations, this book would have been completely impossible.

I also appreciate the reviewers of this text who took time from their own busy lives to review the work and offer significant and thoughtful feedback that has been instrumental to the development of a higher quality book.

Finally, I acknowledge several colleagues who have helped me in many ways with this book: Mike Savoy, Luis Almeida, Husra Gursoy, Brian Beabout, Tom Chermack, Chris Hoadley, Rucha Modak, and Derek Mulenga reviewed pieces of this book and were immensely helpful in developing its intellectual contribution. Thank you.

1

User-Design: The Basics

INTRODUCTION

This book is about user-design; about the idea that we as instructional design (ID) professionals in schools, universities, and organizations of all sorts can sincerely engage users in the creation of their own systems of learning—human learning. User-design is founded on systems theory and thinking and can be concisely defined as an authentic empowerment of a particular set of stakeholders, the users of any innovation, such that they are creating their own systems of human learning. Systems theory and thinking are fundamental for the effective application of human performance and instructional design technologies to organizational and educational change efforts. One of the cornerstones of systemic change is the involvement of all stakeholders in what is termed participatory design or user-design. User-design seeks to produce new systems of human learning and organizational functions to support those systems. It assumes that the users are responsible for design and that those who have traditional design expertise are tasked with facilitating users in the processes of design.

While the value of including the users in the creation of large systems of education and human performance (such as training, computer systems, and curriculum) is apparent, the reality of such inclusive efforts has a history of failure. Meeting the challenge of shifting power dynamics, empowering stakeholders and educating for design must, at some level, fall to the leaders of any dynamic organization. This means that there is clearly a political nature to the actual practice of user-design. This chapter defines user-design and begins to embed it in a systemic change context. By using clear examples and nonexamples of user-design, this chapter brings into sharp focus the meaning and basic ideas associated with user-design. It also starts to link the user-design constructs to instructional design practice.

USER-DESIGN AND THE ROOTS OF ID

Definitions of instructional design practice and Performance Technology (PT) have varied widely through the last decade. From our roots in B. F. Skinner's (1961) behaviorist prescriptions for instruction and Gilbert's Performance Engineering (1978) concepts, we have progressed far to reach the introduction of constructivism (Jonassen, 1991), Performance Technology (Stolovitch & Keeps, 1992), feminist pedagogy (Maher & Tetreault, 1994), situated cognition (Brown & Duguid, 1994) and the learning sciences (Carr-Chellman & Hoadley, 2004). ID can be contrasted with PT, though both share a history and similar approaches. ID focuses on the creation of effective instructional moments where PT may include noninstructional interventions such as incentive systems, organizational development, motivation systems, and strategic alignment. This book is primarily concerned with ID; while PT is dealt with as a separate entity with its own chapter (chapter 8) devoted to examining user-design within PT. Therefore, we concentrate on user-design for ID here.

Examples of the broadening of the field of ID are many. The seminal *Handbook of Human Performance Technology* (Lineberry & Carleton, 1992) points out that diverse, noninstructional interventions should be considered when trying to improve overall human performance in any organization. The recent edition of Dick and Carey's *Systematic Design of Instruction* (2005) includes consideration of issues such as contextual analysis—a critical concept in systemic change. Over the past decade, there has been a shift from a more narrow focus on training and instruction to a broader consideration of various interventions and cultural contexts (e.g., Land & Hannafin, 1996; Henderson, 1996; Gayeski, 1995; Rowland, 1994; Dick & Johnson, 1993). A particularly strong contribution in this widening of the field of design has been Laurillard's text on university teaching (2002) which addresses constructivism as on teaching strategy, but also illustrates the importance of contextual analysis and organizational infrastructure within the design methodology. Most recently, links between instructional design and learning sciences (Carr-Chellman & Hoadley, 2004) have broadened the field of practice.

Despite this expansion, much of the literature in the field continues to focus on efficiency of human learning, instruction, and performance. Our traditional models of ID maintain a sense of linearity and closed

USER-DESIGN: THE BASICS

boundaries. While efficiency seemed an appropriate value in the industrial era, it is becoming increasingly obsolete in the information age where dynamic change in organizations continually forces the integration of new and innovative technologies and processes into the workplace. Because of this dynamism, innovation adoption rates cannot be hampered by less-than-enthusiastic users. Moving through stages of adoption such as those suggested by Rogers (1995) or Bhola (1977) in models of innovation is time consuming to the point that a new process or innovation often comes just on the heels of the adoption of the previously "new" process or innovation. This creates frustration on the part of the user and confusion about his or her place in the broader system. The frequency with which these changes are occurring is exacerbating an already less-than-perfect adoption process, causing anxiety and frustration—which can lead to sabotage of the innovation or a loss of human potential because of high turnover rates. In a way, user-design can be compared to rapid prototyping in that we are trying to move the process more quickly directly into the hands of the users; however, as we shall see, the process of user-design is itself time consuming, ill understood, and creates its own resistance. The best instructional designs are useless without faithful adoption and implementation of the design specifications. This is particularly true of innovations within the design itself. So, we are left with a conundrum—that is, we have great design models that leave us with excellent designs which are only partially and often unfaithfully adopted and implemented via inadequate diffusion models. All of this is, in a sense, a problem of power. We need this book because it helps us to see the ways in which we can overcome this conundrum and create designs that are more likely to engender faithful adoption.

STAKEHOLDER INVOLVEMENT

One way to ameliorate some of the frustration of conducting ID within dynamically changing organizations and bureaucracies is to engage stakeholders in the design of their own systems. Stakeholder engagement is *essential* to the success of design, adoption, and implementation of broad innovations such as new educational systems (Banathy, 1991; Reigeluth, 1993; Jenlink, 1995) or more narrow curricular innovations such as effective uses of video scaffolding technologies: They will all benefit from increased stakeholder engagement.

There are many ways of engaging groups of stakeholders as we design new systems of human learning. The first step in stakeholder engagement is identifying who we should consider stakeholders. The

Handbook of Human Performance Technology (1992) generally says very little about stakeholders, but Lineberry and Carleton (1992, in the Handbook) define stakeholders as "the individuals and groups significantly affected by the organization's performance and results" (p. 238). They identify the major categories of organizational stakeholders including customers, employees, shareholders, suppliers, distributors, and the general public. Using this same definition of stakeholders, we can say that students, teachers, school board members, administration, and community members are the major categories of stakeholders in public school. In general, the literature on stakeholder participation tends toward empowerment, particularly of specific populations (Bauch & Goldring, 1998; Cochran & Dean, 1991; Comer & Haynes, 1991; Delgado-Gaitan, 1991; Midgley & Wood, 1993; Pena, 2000; Romanish, 1993). However, these examples, while they attempt to take stakeholder involvement to the empowerment level, continue to rely on approval and negotiation among leaders and expert designers.

Traditional instructional design has concerned itself primarily with the engagement of stakeholders only as they impact a needs assessment. Generally speaking, in a needs assessment, learners are considered if they happen to be involved in the interview or fact finding process and possibly again only if they happen to be members of a pilot group for formative evaluation. Otherwise, the traditional instructional design paradigm does not include users as designers, but rather as sources of information *for* the design of new instructional solutions.

When stakeholders *are* considered in performance technology, instructional design, or traditional educational reform, methods for identifying them range from leadership alone generating a list of groups and their representatives (the most common) to observing the context and identifying users, stakeholders, and opinion leaders as good sources of information. An additional method is the organization of a group of people concerned with the design project, but not able to participate in the actual design work. This group can then identify candidates for user-design activities and recruit, contact, and confirm their participation (Jenlink, Reigeluth, Carr, & Nelson, 1998). Each of these methods has a variety of tradeoffs; a complete discussion of stakeholder groups and methods for garnering stakeholder participation is offered in Carr (1995b).

But, stakeholder participation is NOT the same as user-design. I introduce it here primarily as a way to link the conversation in user-design with the most familiar touchstones in traditional ID. While stakeholder participation is an important method for gathering support and information for any innovation, including new instructional programs, it is not equivalent to user-design. The critical features which distinguish user-design from other forms of user-input such as stakeholder participation and user-centered design are power, design, and context related. So what, then precisely, is user-design?

USER-DESIGN DEFINED

User-design extends stakeholder involvement beyond mere input to create empowered users who have design and decision-making powers. This is perhaps its most critical feature in terms of differentiating user-design from other forms of user input—that is, the power dynamics change much more substantially in user-design than in any other form of user engagement.

User-design also honors the front line users as the most powerful designers in the system, rather than considering all stakeholders at approximately equivalent levels of influence. In past stakeholder-based approaches, all stakeholders were considered equal and usually were asked to work together through some particular issue in the hopes of reaching consensus or compromise. It is naïve to assume that merely putting all stakeholders into a room together will somehow eliminate or equalize the power dynamics that are already at play in our culture. Thus, the school board member holds more power than the teacher, and the teacher more power than the parent and so forth within a public school arena. And to even complicate matters even more, we cannot simply understand power as attached to simple titles or positions within a system. Indeed, opinion leaders may hold significant power in ways that initially elude the casual observer. We may be surprised to discover a wealth of power residing in the support staff, janitor or lunch lady when we expect to find it in the manager's office. Thus, we need something more, something that actually addresses the underlying natural power dynamics present in all groups of designers and stakeholders and proactively tackles some form of power redistribution. The approach in which stakeholders are more than just "involved" in change and design is often referred to as user-design by systemic change theorists (Banathy, 1991; Reigeluth, 1993; Jenlink, 1995; Carr-Chellman, 1997).

In user-design, as in stakeholder participation, control percolates from the bottom up. Kevin Kelly (1994) describes the problem of control over distributed systems (such as most social systems). He uses the example of bees in a hive to define distributed systems, but generalizes the point to various social systems including organizations, the internet, and learning systems:

> When everything is connected to everything in a distributed network, everything happens at once. When everything happens at once, wide and fast moving problems simply route around any central authority. Therefore

> overall governance must arise from the most humble interdependent acts done locally in parallel, and not from a central command. A mob *can* steer itself, and in the territory of rapid, massive, and heterogeneous change, only a mob can steer. To get something from nothing, control must rest at the bottom within simplicity. (p. 469)

Stemming from Scandinavian approaches to software interface design, user-design is an approach that has been applied to the design of computer systems in which "the people destined to *use* the system play a critical role in *designing* it" (Schuler & Namioka, 1993, p. xi). So, we now have a working definition for user-design; however, knowing the definition of user-design is only really useful if we understand *how* to use it and what it is within the *context* of systemic change. We will address the how question further in chapter three and the contextual question in more detail in chapter two. However, a brief discussion of the systemic change context is important in a clear and complete definition of user-design, so it is included here. A more in-depth discussion of systemic change context is in Chapter 7.

THE IMPACT OF USER-DESIGN ON ID/PT

One of the cornerstone values of systemic change and UD is the fundamental assumption that users must negotiate change with leaders on an equal basis. Because of this, the systems designer role shifts rather dramatically. First, the systems designer in either the ID or PT context must address issues of power and resistance, working with leaders to help them see the hazards of leaving the users out. In addition, the instructional designer or performance technologist must *work with* users to create ideal systems of human learning. Thus, there is little or no space in user-design for expertism (Carr-Chellman, in press); instead, users are empowered to create their own visions apart from the agenda of the designer. An aboriginal woman, Lila Watson, once said, "If you've come to help, you're wasting your time; but if you've come because your liberation is bound up with mine, then let's work together." This is a common saying among systemic change advocates and it is one of the fundamental values which can be difficult to live by because it calls into question the validity of our presence in a system as an expert, thereby calling into question our training, our pay scales, even our very purpose. This shift from expert system designer to design facilitator exemplifies the desire of the systemic change movement to realign power distributions and alter the dominant paradigms of top-down change. I discuss this change for instructional designers in more depth in Chapter 3.

USER-DESIGN: THE BASICS

User-design represents a dramatic shift in power dynamics from traditional approaches. In traditional diffusion of innovations, the reformer analyzes, creates, and negotiates; the leaders initiate, approve, and decide. Unfortunately, the users are left to accept or reject the innovation, and much literature has focused on better and better ways to encourage adoption or compliance from the end users. This approach, however, does not give weight to indigenous knowledge and does not recognize that smart stakeholders usually realize that they are being, in large part, controlled or manipulated by the negotiated agenda of the designer and the leader. Typically, those products or processes which are truly designed by users tend to build ownership among users and create a significantly different adoption process than is typical of more manipulative models of innovation adoption. Perhaps the most well-known diffusion theorist is Evrett Rogers, author of *Diffusion of Innovations* (1995). Rogers's approach, or the "colonial" approach to design and diffusion has been critiqued because of the disempowerment of users and the lack of respect afforded indigenous knowledge (Carmen, 1990; Yapa, 1996a, 1996b). This traditional approach is deficient in terms of the robustness necessary given the variability of many current contexts (Larsen & McGuire, 1998). Thus, in user-design, actions such as initiation, approval, rejection, design, and decision making are negotiated among the users, designers, and leaders.

Certainly, the recent passing of Ev Rogers saddens us all. And his work is significant and not to be minimized by this critique of colonial models of diffusion. In fact, he gave us a language to use that helps us to really understand diffusion of innovations when we had nothing. However, the approach is very much top-down. It was interesting to me that even his obituary, written, presumably by one of his students, Kim McCormick of the University of North Florida, shows this "light bringer" attitude:

> Please join me in honoring Everett Rogers by pursuing his dreams and goals; to teach and care about students by doing, to conduct research that matters to people, and to do everything we can to make the communities in which we live mirror the wealth of knowledge we possess.

It's a lovely eulogy, and I agree that we should care about students and do research that matters; it's that last part that bothers me, the sense that we possess a wealth of knowledge that we need to bring to our communities. It is this method in which power is bestowed upon experts and reserved from those who posses unrecognized wealths of indigenous knowledge.

This bottom-up approach to the design of technology, curriculum, incentive systems, or organizations as a whole can be quite intimidating for both systems designers and users. As an example, if a technology curriculum is being developed, the traditional approach would be to conduct a needs assessment, identify the technology learning needs of the various grade levels, and prescribe a curriculum based on these findings which would then be designed and delivered to teachers and learners as expert advice often in the form of pre-packaged, vendor-delivered curricular materials. User-design of the same curriculum would require full engagement of multiple users in the understanding of the localized, current, and ideal situation as they create their future learning. Users would engage in a number of experiences to learn more about possibilities for future technology applications to learning. Based on this, the design facilitator would then work with a team of users to help them create a curriculum which comes closer to their idealized vision.

This process will appear less efficient, and indeed working by committee can often produce frustration and slow progress. In addition, such a model of stakeholder-based design creates confusion over issues of power and control. If we cannot conduct a needs assessment, find a problem, and propose a prescriptive solution, what do we have to offer? The answer to this question is, we have a process (a soft technology, if you will) by which sustained positive change can be accomplished to meet the needs of a given community of learners. There are tradeoffs in the application of user-design. While the process is powerful, it is also less clearly marketable than traditional instructional design solutions based on design expertise precisely because it is more subtle and complex. I was recently involved in a story for *Prime Time Live* the television news show. I found that anything that was at all complex or subtle, confusing or conflicting was cut from the show. Most people prefer their lives to have clear order and things like user-design, while empowering and exciting, can mess that order up quite a lot. Just the same, I argue that the long term benefits in terms of user ownership, improved designs, and authentic adoption cycles clearly outweigh the short term drawbacks of inefficiency and resistance.

USER CENTERED DESIGN

It is important to distinguish user-design from user *centered* design and other forms of user input. In the former, users are engaged in the actual creation of their own systems in negotiation with leaders and designers. In the latter, users are considered central to the design specifications; however, design control remains firmly in the hands of professional designers, and approval power remains with leadership. This could be compared with representation as in our political structure. While well-intentioned, representative politics still create a sense of apathy and disenfranchisement among the general populace. In this same way, user-centered design or learner-centered design (e.g., Urbanski, 1995; Holloway, 1972; Norman & Draper, 1986; Sugar, 1995, 2000; Dillon, 1995) differs from user-design primarily in the location of decision-making power. User-centered design has become very popular among those who seek to understand and even harness the knowledge that is indigenous to users. For example, there is even a specific International Organization for Standardization (ISO) standard which addresses the need to consult users in the creation of interactive systems (see http://www.usabilitynet.org/tools/13407stds.htm).

However, user-design is *not* the practice of increasing "user involvement in acquiring, maintaining and manipulating essential institutional data" for the purpose of "incorporating user input into systems design and development" (Hurley, 1980). It is also *not* user-based (Abels, 1997) design in which focus groups and questionnaires are used to gather user perspectives for application to the design of systems. All of these are admirable strategies for both soliciting input into the design of a new system from users and garnering user support for the newly designed system. But, none of these approaches really takes the power dynamics and significantly changes what we do. These approaches do not attempt to wrest from the hands of the designers the actual tools or knowledge of design procedures and to pass this largely common sense knowledge on to the users. True user-design, as messy, inefficient, overwhelming, difficult, contentious and perturbing as it may be to the system, goes beyond mere consultation to elevate the user to the role of a designer. While the emphasis on accounting for the needs of the users is primary in user centered design models, the actual process does not look all that different from current traditional models of design because no shift of power has occurred. In both cases (user-design and user centered design), however, leaders must be convinced of the importance of user perspectives and needs and will have to lend their support to efforts in order to see effective implementation of either approach.

Despite the differences between user-design and user-centered design, there is a tie to the cognitive psychology literature here with regard to other similar learner-centered theories of learning such as constructivism (Duffy & Jonassen, 1991), problem-based learning (Barrows, 1986; Stepien & Gallagher, 1993) and learner controlled environments (Land & Hannafin, 1996). Broadly based on Deweyan (Garrison, 1995) and Vygotskian (Davydov, 1995) understandings of human learning, these recently emergent practices of instructional design are examples of user-design at the classroom (or micro) level. By shifting the power in the classroom away from the teacher into the hands of the learners, there is a commonality with user-design and in some cases, the ultimate goal or epitome of the radical constructivist (Glasersfeld, 1995) or deconstructivist classroom would be the ideal of user-design (learner-design). Further discussion of learner centered performance technology is taken up in chapter 8.

The ideal learning experience, then would take the form of a question to the learner: "What do you want to do today? What do you want to learn today?—O.K., let's create a way together for you to learn that." However, as it currently is actualized in instructional design, constructivism (e.g., Duffy & Jonassen, 1991), learner-centered, and learner-controlled environments (e.g., Land & Hannafin, 1996) typically do leave many design decisions up to the designer rather than the learner. We have not yet released ourselves from the idea that someone else knows best what learners should learn and that no one would be at all interested in learning the hard stuff if left to their own devices. In fact, the standards based movement and the No Child Left Behind legislation of this educational reform era substantially erodes any advances the learner-centered movement had made in the past decade. Instead, as has been the case for years in the past, the designer or teacher provides the learners with a case for case-based learning (Barrows, 1986), a problem for problem-based learning, or a context into which situated cognition (Cognition and Technology Group at Vanderbilt; 1993) is enclosed. This belies a certain industrial model of education and learning where a boss pretty much still knows what is best for the laborers.

Thus, while there is a strong tie between these areas, they are not precisely the same as user-design and, therefore are appropriate for different applications. User centered or learner centered design is appropriate where some level of designer power needs to be maintained, while user-design is best considered when that expertise is problematic because of anticipated implementation problems.

IMPLICATIONS FOR DESIGN

While all of the implications of the above design methods cannot be anticipated, it is clear that a very different landscape of instructional, interface, and organizational design emerges from the values and processes embraced by systemic change and user-design. I provide the following questions (See Table 1) as an initial attempt to clarify what sorts of issues arise when attempting to design under the rubric of user-design. Of course it is important first for the designer to ask

USER-DESIGN: THE BASICS

TABLE 1
Is User Design Appropriate?

Questions to Help Determine If User Design Is Appropriate

Is leadership secure? Are they able to consider power sharing?
Are ready-made solutions unlikely to be adopted? What is the history of change in the organization?
Is user resistance high?
Is the problem Systemic?
Is there sufficient time (as a resource and in actual deadlines) for the protracted user design process?
Are users engaged in their organization?
Are users already equipped or motivated to gain design skills?
Are better processes or products needed?

him/herself whether or not user-design is the appropriate approach. If user-design *is* the desired method for the creation of user-based products of any sort, the designer must first situate the design work in a systemic context (see Chapter 7) and then implement a given user-design technique (see Chapter 3). Not all of these components are necessary for all user-design projects, but these questions will help the designer to focus his/her user-design efforts.

User-design is certainly not appropriate for all situations. User-design is generally more appropriate when expertise may not be easily accepted by the users, when there is some source of resistance (e.g., Will I lose my job or identity?) and when the problem is systemic. Very narrow problems, for which expertise is available and appropriate, are not likely candidates for user-design. In addition, if a community is not capable or motivated to build their capacities in the process of design, then user-design is not likely to be any more successful than traditional methods and may prove a good bit more frustrating and inefficient.

There are a wide variety of tradeoffs associated with the process of user-design. In particular, a protracted design timeline and accompanying questions of efficiency are limitations to the current application of user-design in schools and workplaces. In addition, those undertaking user-design need to have specialized skills in leadership and facilitation. I have began to outline the advantages and disadvantages of user-design in Table 2, but I'm sure you'll be able to brainstorm more as you learn about user-design and apply it to your own particular design practice.

Naturally, while user-design *seeks* to create more powerful designs, there is always the possibility that users will create something that is

TABLE 2
Advantages and Disadvantages of the User Design Approach

User Design Advantages	User Design Disadvantages
Faster adoption rates	Long timeline and time intensive
Better processes and products	Lack of efficiency
Continuous improvement of entire system	Specific skills in user design facilitation and practice needed
Rewards workers and users with more interesting and inclusive work	Some designs resulting from UD may be unworkable or unrealistic
Significant financial savings during implementation and adoption	Bad or harmful ideas may result
Integrated systemic approach to change and improvement	
Creates agile organization with iterative trials of new products and processes	
More democratic work environment	

unworkable or simply a bad idea that doesn't function well in the setting. However, bad ideas are often in the eye of the beholder rather than some absolute measure. Tradeoffs may be necessary in order to initiate a continual process which empowers users to create their own systems. If negotiation and additional information does not help to mitigate this situation, and if the design facilitator finds the community direction to be untenable or abhorrent, the design facilitator may have no choice but to withdraw from the context. This does not imply that users may not continue along their chosen course, only that the designer, having made a moral, ethical or rational choice, no longer continues to enable the users to create their system. Of course, errors are also possible with expert design; and where idealized design is seen as a continuous process, initial errors or failures can be seen as learning experiences rather than mistakes to be regretted.

CHAPTER SUMMARY AND THOUGHT QUESTIONS

This chapter has examined some of the recent shifts in instructional design theories which embrace more open and permeable boundaries for the practice of instructional design. It has distinguished user-design from stakeholder participation, user-centered design, learner-centered learning, and more traditional forms of engaging learners. It has offered a clear definition of user-design and has embedded it within the systemic change context. The impact of user-design on the future practice of instructional design and the implications of user-design on instructional design have been drawn.

Systemic change, user-design and leadership combine to form a powerful approach to the creation of new systems of human learning. Understanding the organization from a global perspective and understanding change theories along with design theories creates a fertile ground for engaging in the methods of user-design. It is only through such holistic approaches that we will see the next transcendence of instructional design into substantial systemic change technologies. While focused, traditional designs are appropriate for some cases, many of our best designs suffer for lack of appropriate implementation and adoption in real world settings. This dilemma may be transcended through systemic approaches to design and powerful inclusion of the users in design processes. This shift can combine transformational leadership with responsible user level design activities to create shared visions, strong performance motivation, and effective communication practices. Surpassing the current expert-driven change processes to engage truly shared decision making can be frightening for experts and leaders alike. It is my challenge (and perhaps naïve hope) to create a space where users, designers, and leaders can all find power, comfort, responsibility, and satisfaction in the dynamic system that truly encourages learning.

> In your own words, describe the basic concepts of UD.
> How is UD related to other innovative ID approaches?
> What is the most radical aspect of UD in your view?
> What do you think the future of UD is likely to be?
> How would you define UD in one sentence to a fellow instructional designer?
> How would you define UD in one paragraph to your parent or spouse?

2

History and Research[1]

INTRODUCTION

The purpose of this chapter is to describe the history and current state of research and theory in user-design and connect this foundation to organizational change and instructional design. In recent decades the importance of the worker, learner, and user has increased considerably, and attention paid to user participation in organizations has been on the rise. However, one closely related area of research and theory, that of user-design, and its potential contributions to engaging whole communities in the design of their own systems has not been carefully explicated. This has created situations where top-down mandated changes such as the No Child Left Behind January 2002 legislation, Bush's education agenda, has spurred a backlash among users, particularly teachers who call the movement, "No Teacher Left Standing." Informing ourselves of the history and research foundations associated with user-design will establish working knowledge sufficient to begin engaging in the processes of user-design that follow in chapter 3.

Traditional organizational change efforts have often taken the form of new initiatives being diffused by select leaders or policy decisions most often made at the management level. Because of this traditional innovation flow, frontline users in organizations are often faced with acceptance or rejection as their only options in response to innovations.

---

[1] An earlier version of this chapter was published as Carr-Chellman and Savoy (2004). My appreciation goes to Mike Savoy who collected the information that made this chapter possible and collaborated with me in the initial stages of clarifying the history and research traditions that contribute to user design.

HISTORY AND RESEARCH 15

This has led to certain innovations, which are rejected by the system or ineffectively implemented. Experts and practitioners, being frustrated with the lack of relevant useful research results, have engaged collaborative efforts such as organizational design and strategic planning which tend to be less systemic in nature.

This chapter identifies a typology of user-design and ties user-design to a number of foundations and related disciplines including learner centered design; human–computer interface; emancipatory design; Scandinavian user-design; traditional stakeholder participation and, of course, systems design/systems theory. Within the discussion of each of these theoretical foundations, connection is drawn to the instructional design practitioner in order to suggest ways in which this research can be used effectively by those who wish to engage in change facilitation.

TYPES OF USER PARTICIPATION

Noyes and Baber (1999) define *user* as the "human component" of design. However, this definition gives no detail as to who actually uses and/or benefits from the designed product or process. It also does not define the level of knowledge these users possess. User-participation occurs in various ways depending on the context, participants, resources, and intentionality with which user-engagement is proscribed. There is much confusion about the differences between *levels* of user participation (i.e., user-centered, learner-centered, student-centered, user-design, etc.). While some researchers make no distinction between sublevels of user-centered design (Sugar, 2001), others distinguish between user-centered and user-participation (Salvo, 2001), and some clearly define various levels of user-design (Schulze, 2001).

Regardless of the sub-level of UD, the designer and/or leader, typically determines the extent to which users are engaged in the creation of their own systems. Thus, grassroots movements (Jackson, 1993; Merrifield, 1993; Olson, 1990) are infrequently possible. Because the leader typically maintains power in most organizational contexts and is too often uncomfortable giving it up, true user-participation requires a different perspective on organizational structure and radically different communication systems.

User-Centered Design

In the first chapter, I lumped user-centered design and learner-centered design together. However, for the purposes of clarity, I now divide user centered design into two broad groups of models and corresponding literature. The first is primarily concerned with learners, while the second is more concerned with end-users. In general, the learners are primarily engaged in structured or semi-structured learning experiences, while end-users are engaged in tool use. I explore learner-centered design first, and then user-centered tool design second, and distinguish between user centered tool design for human computer interfaces and library media tools.

Learner-Centered Design. In traditional learning situations, consideration for the learner is usually afforded by needs assessments, pretests, learner analyses, and sensitivity to individual learner differences, however, the learner does not actually have a *say* in what, when, how and to what extent they learn. Learner-centered design differs from other types of user-design in that the focus is on learning and pedagogy rather than tool use as is the case in Human Computer Interface design. Learner centered design emanates from learner-centered psychological principles (American Psychologial Association, 1993), particularly those associated with metacognition; cognition and affective, developmental and social psychology. Perhaps the most well known theorist to extend learner-centered principles from situated cognition (Brown & Duguid, 1994), constructivism (Duffy, Lowyck, & Jonassen, 1993; Jonassen, 1999), and systems theories (Banathy, 1973, 1996; Senge, 1990) was Barbara McCombs (2001; McCombs & Whistler, 1997).

Learner-centered design encourages active collaboration and engaging learners "as active participants in the generation of learning plans" Wagner & McCombs, 1995, p. 33). There is little discussion of the shifts in power that are necessary to engage learners in substantive ways in the current literature on learner centered research. All things should focus on successful outcomes for the learner rather than other extraneous concerns such as leader desires, contextual cues, or resource limitations. This position argues that learners ought to be afforded a serious opportunity to influence their own learning (Wagner & McCombs, 1995), and this approach allows learners to take an active role in their own learning.

In terms of empirical studies in the area of student centered or learner centered environments, there are a number of strands of research. The seminal author in this area, as I've already mentioned, is McCombs. She has four studies in particular which indicate positive findings for learner-centered classrooms (Daniels, Kalkman, & McCombs, 2001; McCombs, 2001, McCombs & Quiat, 2000; Weinberger & McCombs 2001). Salisbusy-Glennon, Gorrell, Sanders, Boyd, & Kamen examined the implementation of learner-centered philosophies. This study found that certain self-regulation strategies such as seeking, organizing, and transforming information were used more often than memorizing, self-evaluation, and record-keeping for those in the learner centered situation. Taken together, the studies on learner centered learning are hopeful and encouraging, indicating that in proper contexts, learner centered approaches can work well. Turning learning entirely or primarily over to the learner is still not a part of this dialogue, however.

Tool Usage

The application of user-design princples to the creation of artifacts and tools for human productivity has primarily been utilized in two areas: Human Computer Interfaces Design (Norman & Draper, 1986) and library media use (Fidel, 1994; Morris, 1994; Wilson, 1995). Where organizational change leaders are concerned, this represents one of the richest resources of understanding how stakeholders and users can be authentically engaged in the creation and design of their own systems of human learning.

Human Computer Interfaces (HCI). User-design for the creation of computer based tools has focused more on the "how-to" (work with users) than learner-centered models. However, most user-centered design literature maintains power in the hands of designers. Sugar & Boling (1995) advocate for early user engagement and iterative processes to most effectively represent user desires in final products. Norman's edited text on *User-centered Systems Design* seems to focus primarily on how best to engage users in the creation of their own systems. For example, the chapter by Riley (1986) focuses on those fundamental understandings that users need in order to have to engage designers. Certainly to the extent that Participatory Design (Schuler & Namioka, 1993; Caroll & Rosson, 1992) is related to human computer interfaces literatures, this is perhaps the closest kin to user-design; however, the power shifts are not the central focus in the HCI literatures as it is within the field of user-design.

Media Usage. In the case of library media usage, user centered design has primarily meant conducting user surveys in order to better design the library resources and systems for patrons' use (DeCandido, 1997; Rockman, 1980; Wilson & Arp, 1995). Starting in the mid 1980s, information retrieval methods were increasingly influenced by advancements in technology (McCandless et al., 1985). These advancements coincided with library/media center users wanting information systems that were characterized by easy, adaptive, user friendly interfaces and navigation tools (Payette & Rieger, 1998). These systems needed to be suitable for novice technology users, handicapped individuals, as well as the typical library patron. This stream of research is, however, only tangentially related to what user centered design for instructional purposes has meant. User centered design for instructional systems has been more closely aligned with Donald Norman's understanding of design for human computer interfaces discussed earlier.

Emancipatory Design

Emancipatory design models take the mission of empowerment beyond user-design. The emancipatory reform team hopes to inspire transformation, to alter some significant, and often historically intractable aspect of society. The goal of emancipatory design is more to create change and to vest the users in organizational outcomes than it is to concern itself with the more mundane task of creating a functional organizational or human learning system. Emancipation as it is applied to action in the form of education or design emanates from Paolo Freire's work with Brazilian and Chilean illiteracy. It was Freire's contention that knowledge collaboratively constructed is the key to changes in practice. He asserted that research itself is a project of social change, and his understandings were instantiated as the popular education movement (Morrow & Torres, 2002) in South America during the 1960s and 1970s (I. C. Carr, 1990; Gerhardt, 1986; Melo & Benavente, 1978).

Suzanne Damarin has done a very good job of making the connection between emancipatory pedagogy and instructional technology. (2001). She points out that the traditional system of education assumes that the learner is not an active participant in the construction of their learning environments, but rather are seen as repositories of facts. Damarin (2001) explains that emancipatory pedagogy, in contrast, is about shifts in power:

> Also common to these theorists is the notion that teachers and their authority within the classroom must be *decentered* in order that the knowledge students bring to the classroom from their homes and communities be honored as grounds on and from which to build new knowledge. (p. 17)

Emancipatory ideas, then, have been used in education and research but more rarely in design or organizational change activities. Designers have been more interested in the creation of appropriate, functional, cost-effective artifacts than in the ways those artifacts may serve to free oppressed populations. In fact, because return on investment is the primary motivator of most for-profit organizations within the U.S., capitalism has prevented engagement in emancipatory design as it essentially threatens the very existence of corporate America. While most emancipatory models have been applied to people of poverty, many users need emancipation within the for-profit corporation today. This emancipation came in the form of unionization in the past, but that resulted, at times, in representative democracy, contentious leadership relations, and reactive concerns rather than forward looking design practice. The knowledge management theorists notwithstanding, corporations are less than interested in true transparency of knowledge sharing. It is still the case that information is power, and budgets are kept close to the vest among leaders.

In both the case of user-design and emancipatory design, the payoffs for the organization are tertiary rather than primary or even secondary. That is, the rewards are usually more to the users than to the leaders. There exists little in the way of literature on what I am calling emancipatory design. This is quite possibly because it brings power issues into sharp focus as they conflict between users, designers, and leaders. It may be seen as highly impractical particularly in for-profit organizations where there is often clear conflict between leaders and frontline workers, management and labor, or unions. There may be rather more support for an emancipatory approach in not-for-profit and NGOs. In the case of emancipatory design, the user is in charge; their power their indigenous knowledge is *more* powerful and respected than that of the expert. Because of this positionality, emancipatory design draws heavily on critical theories which have as their central focus who benefits and who is disempowered by any innovation, policy, or product.

WHERE DOES UD COME FROM?

User-design research and theory is drawn from a wide variety of philosophical stances, research traditions, and theoretical perspectives. Considerable work has already been done to explicate many of these foundations, though their relationship and contribution to the aims of user-design have not been drawn. The purpose of this section is to briefly describe the foundations of user-design models and to indicate their relationships and contributions to the discipline of user-design broadly defined. There are three primary foundations for user-design: the Scandinavian tradition, stakeholder participation research, and systems design/systems theory. I have touched on these in the first chapter as they contribute to the definition of user-design. Here we explore the research in each of these foundational fields as they contribute to a deeper understanding of user-design.

Scandinavian User-Design

As I mentioned briefly in Chapter 1, Scandinavian user-design has been primarily applied to the creation of user-friendly software interfaces. As laid out by Schuler & Namioka (1993), user-design within the Scandinavian tradition is an opportunity for those users who will ultimately come in closest contact with the final interface to play an essential role in the design of that interface.

Scandinavian researchers have a long history of user-centered, user-design, and emancipatory design literatures. Their active involvement of users in systems development traces back two decades or more (Bansler, 1989; Bødker, 1996). Schuler and Namioka (1993) set out the Scandinavian roots of user-design as they apply to the creation of information technology interfaces. Scandinavian participatory design research not only focuses on improved product development as a result of user participation, but also the political structure between management and labor. However, the Scandinavian work force is characterized by high education levels, strong unions, and laws regulating management/union relations.

Certainly, Schuler & Namioka's 1993 text, *Participatory Design: Principles and Practices,* is one of the seminal texts in the field. While their primary emphasis is on human compter interfaces tool use, many of the ideas are useful for stakeholder based educational reform although, to the best of my knowledge, these ideas and research have not been widely disseminated among American public schools or organizations.

Perhaps one of the best reviews of research in Scandinavian user-design is offered by Bjerknes & Bratteteig, who focus primarily on the relationship between users and administrators and the inherent contextual issues that are defined by user-design. They trace the roots of user-design back to a series of research projects involving trade unions and define two strategic perspectives—conflict or harmony. In this case, Bjerknes & Bratteteig suggest that the conflict perspective recognizes the inherent contextual conflict between users–laborers and administrations–employers and calls on the researcher to work on behalf of the less powerful (i.e., the users, laborers, or employees) to empower them. The harmony perspective suggests that all are working for the betterment of the organization and so all interests are aligned rather than being oppositional. Growing out of this review is a careful explication of the relationship among critical theory, democracy, and social change.

Most Scandinavian user-design research has focused on HCI and tool design, though there has been some literature which looks specifically at work contexts and the design of jobs for democracy (Elden, 1979). In fact, the Scandinavian cultures take the user-design process so

HISTORY AND RESEARCH 21

seriously that in some cases, such as Denmark employment law and the Norwegian Worker Protection and Working Environment Act, it has been legislated (AML, 1977; Norrbom, 2001; Otten, 1991). As I mentioned earlier, the relationship between Scandinavian UD research and the application of UD is still quite limited. Very few organizational change theorists or school change agents are using (I suspect they may be unaware of) the Scandinavian HCI research. However, I believe there is a natural extension from user-design to stakeholder participation and that the user-design literature can therefore inform our practices with systemic change. While I recognize the inherent power shifts necessary for effective user-design (A. A. Carr, 1997), Scandinavian researchers tend more toward a deconstruction of the social context of work in an effort to uncover the inherent conflicts as obstacles to user engagement.

Stakeholder Participation

User-design extends stakeholder involvement beyond mere input to create empowered users who have design and decision-making powers. However, as mentioned in chapter 1, the links to stakeholder participation literature and research are strong. Enacting substantive change requires more than a mere open invitation to stakeholders to participate. Each unique situation determines who the users are, and each user has a different experience and knowledge level. In the most effective cases of both user-design and stakeholder participation, control percolates from the bottom up. Grassroots movements, while rare, are perhaps our strongest cases of true user-design built on the foundation of effective stakeholder participation. As described in Chapter 1, Kevin Kelly 1994 in his text, *Wired*, asserts that within social systems where everything is connected to everything else (a lesson we are rapidly learning firsthand in the global economy of today), things happen quickly and "simply route around any central authority" (p. 469). Stakeholder researchers have, for some time, realized that stakeholder participation is one way to stem the ineffective implementation of innovations due to this "routing around."

User participation, as with stakeholder participation, becomes more complex as the size of the system involved increases. For example, stakeholder research informs us that successful participation requires multi-level and multi-stake participation (Daresh, 1992; Stevenson & Pellicer, 1992). This means that for a complex social system such as an organization, stakeholders ranging from leaders to workers, from trustees to clients must engage in the design or participation processes at many levels including front line, management, whole organization, even the broader marketplace. Stakeholder participation theory and research offers the user-designer a number of similar lessons about effective implementation of empowering methods (Berube, 1970; Cooper, 1992; Davies, 1981; Epstein, 1997; Fantini, Gittell, & Magat, 1970; Sarason, 1995). In general, the findings of the empirical research studies in stakeholder participation tend to point to positive, effective uses of stakeholders as sources of information, but this is more in the user-input vein than the user-design vein of change. That is, few cases of real decision-making power accorded to stakeholders are present in the current research on stakeholder participation. Nevertheless, this promising line of research indicating positive results in using stakeholder participation can lead us to hope for similar or more powerful results from user-design approaches to ID.

Systems Design/Systems Theory

As I stated earlier, user-design is intimately linked with systems design and systems theory. This is the third foundational literature that we will consider as part of the heritage of user-design. In fact, systems thinking is one of the cornerstone foundations of user-design (Ming-fen, 2000). There are many reasons for this connection. First, systems theory shares certain values and beliefs in common with user-design such as holism, interconnectedness, interdependency, and respect for indigenous knowledge. Another reason why user-design and systems design are linked is because truly effective utilization of user participation requires that users understand how the system *they* are a part of works. For example, a support staff team charged with making decisions about a new information system will have a difficult time making realistic recommendations unless they fully understand the interconnections of the information system with other systems such as communications, finance, and even governance.

Educational systems theories (Banathy, 1991; Jenlink, 1995; Reigeluth & Garfinkle, 1992), specifically, share many of the fundamental values of UD such as the importance of engaging users in the design of their own systems rather than devising ways to convince them to accept experts' designs. Social systems theories have always embraced the "softer" side of understanding wholes (Checkland, 1999). Unlike their counterparts in hard systems theories (Rapoport, 19986; von Bertalanffy, 1968), social systems theorists have fought long and arduous battles to forward the agenda of a valued systems theory—one that values humans placed in a post-Cartesian context.

While some of the ideas of hard systems theories are applicable to user-design (e.g., living systems [Miller, 1978], liberating systems theory

[Flood, 1990], and dynamical systems [Jantsch, 1980; Prigogine & Stengers, 1984]), most of the commonalities are drawn from social systems or soft systems methodologies (Checkland, 1999; Churchman, 1968). These methodologies address issues associated with human activity systems and problems of change, design, and adoption implementation in instrumental ways for the user-designer.

An example of the ways in which systems theories enlighten a specific communication problem may illustrate this connection. Dissemination of information across open systems boundaries calls for translations from researcher or design expert to user or stakeholder. Understanding system boundaries, permeability, interdependencies, and interconnections among the subsystems and suprasystems in question is necessary for effective communication in the case of a researcher trying to communicate her findings to a practitioner based audience. User participation reduces the need for efficient and accurate translations by one source in order for the relevant information to be carried throughout the system.

CHAPTER SUMMARY AND THOUGHT QUESTIONS

This chapter has reviewed the basic theoretical foundations and related disciplines surrounding and supporting user-design. I hope that this chapter has helped you to contextualize UD within a broader research literature as well as to understand clearly that which has come before and contributes to the current understanding of user-design today. I am also hopeful that you will find the user-design literature helpful in thinking about how schools, universities, and all sorts of organizations can engage in new ways of doing design for human learning. There are many examples of ways in which learning about user-design, emancipatory user engagement, and even HCI design can help the instructional designer realize the ultimate goal of improved organizations. For example, lessons from the Scandinavian literature can be used to inform leaders as they embark on worker advisory team formation and engagement. The instructional designer can use these ideas and this research to justify the importance of user-design approaches which may be initially perceived as inefficient. Users and front line workers can use this literature base to encourage others to join them in collective actions.

There are, naturally, a number of obstacles that arise during such a project. Many of these have been well documented elsewhere, but they include lack of time, lack of money, lack of interest, and lack of power. In addition, some of the warnings among the user-design community may be well heeded by the systemic change practitioner.

The primary obstacle is no doubt the problem of power and the reticence of the powerful to engage users in decision making. The tendency to categorize all users into a monolithic group possessing a single set of characteristics, desires, and stakes in the outcome of organizational change is another important obstacle. From the HCI literature, we learn about obstacles such as user motivation, apathy, user identification, communication, value conflicts, user access, obtaining user feedback, and implementing user feedback effectively (Grudin, 1993). Thus, the foundations of user-design can be an important source of information for those engaging in organizational change efforts in many ways from strategies to obstacles. While the theories and research in UD and user-design related literatures are rich, broad, and varied, careful consideration of user-design literature can only strengthen our attempts to build stronger, more flexible, and agile organizations.

> What in your view is the most important type of user participation? Why?
>
> What did you find most important in terms of the research base for UD? Why was that the most important learning to you?
>
> What in your view is the most important of the three foundational literatures in UD? Why?
>
> What do you consider the most important thing that we can learn from the past research in UD?

3

Tools for User-Design

DEFINING TOOLS

As with any form of design, the tools that are employed are vital to the effectiveness of the design process and resultant products. The tools are the things that we use to help us actually *do* design—create. Some would call them methods, others means; in this case, precise language is not as important as understanding the basic underlying concepts of certain tools *and* a clear understanding that tools are *not* value-neutral. In fact, all tools do express certain values (Winner, 1986; Postman, 1992; Heidegger, 1977), and these values may be more or less consistent with the underlying values of the design model that has been selected. Thus, we can imagine that certain tools work well in constructivist design models, and others will not work so well in that environment, while still other tools may be well-suited to creation within a behaviorist design model. The key is to deeply understand that design models all have inherent values expressed within them, and that the tools we use to make those design models into real products also express certain values. Mixing up the underlying values can be very confusing as a designer, so it is essential that we manage to keep the models and tools more or less consistent within the values framework.

Once a clear understanding of the values of both the model and the chosen tools has been accomplished, selecting appropriate tools can be overwhelming. In the case of user-design, some earlier work (A. A. Carr, 1997) lays out three possible tools—ethnography (Adler & Adler, 1994; Fontana & Frey, 1994), cooperative design (Bødker, Gronbaek, & Kyng, 1993), and action-based UD (Park, Brydon-Miller, Hall, & Jackson, 1993) (modeled after action research). These are three approaches, tools that are definitely appropriate within the values framework of user-design. I'd like to explore these three tools and add two more to our toolkit: scenario planning and design-based research. I believe that these two additional approaches can be very helpful in thinking through the processes of user-design and that they can help us to actually engage users in the creation of their own systems of human learning.

Remember that identifying and engaging stakeholders or users is only a *part* of the user-design process. It is important to recall that user-design is *not* about trying to control these groups so that they present the least amount of resistance to change, but rather engaging them in the difficult task of negotiating new systems of human learning. There are several examples of attempts at this sort of empowerment such as total quality management and site-based decision making (e.g., Daresh, 1992); however, these attempts have typically met with less than successful results, primarily because of lack of power shifts (Sarason, 1995) and inattention to culture and system dynamics.

All of the tools considered in this chapter focus on system dynamics and shifts in power that are necessary to truly engage in user-design. All five of these tools share commonalities including some period of time spent observing, interviewing, and otherwise engaging users in conversation (Bohm, 1990) around the system of interest. All methods are built on trusting relationships with users, which requires *authentic* engagement with users in such a way that their ideas are integrated into the design. All methods are embedded in the systemic change context being grounded in holism and systems design and are motivated by an understanding that engaging users not only produces "buy-in," which in turn eases adoption, but also that products created in collaboration with users are *better* products. The first two methods are outlined in detail in Schuler and Namioka's (1993) text on Participatory Design. The third is drawn from many sources on action research, and in particular, *Voices of Change* (Park, Hall & Jackson, 1993). The fourth, design-based experiments (Hoadley, 2004), may initially seem less in line with user-design's values; however, taking the basics of design based experiments and casting them in a user-design light, it is clear that design based research can be a flexible tool to assist in user-design as well as more traditional innovation implementation research. Finally, scenario planning (Chermack, Lynham, & Ruona, 2001) has been examined from narrow to broad scopes—that is from scenario use in instruction through scenario use for community planning. Our interest is much closer to the latter and thus I focus on educational planning with scenarios in user-design in that section of this chapter.

I believe it is very important to understand how each of these tools passes a user-design "litmus test" (i.e., not all potential tools will be

TABLE 3
Criteria for User-Design Tools

Variable	Ethnography	Coop Design	Action Research	Design Based	Scenario
Empowerment	✓	✓	✓		
Broad engagement			✓	✓	✓
Power shifts		✓	✓		
Time spent engaged	✓		✓	✓	
Holism	✓			✓	✓
Respect indigenous knowledge	✓	✓	✓		✓
Non-expert approach okay	✓	✓	✓		
Design oriented		✓		✓	✓
Change seeking			✓		✓

sufficiently consonant with the underlying value systems of user-design to make it a potential match). But, as you think about these tools and as you encounter other potential tools in your own reading, it is important to figure out how to decide if a potential theory or tool might work in user-design or if it is totally out of synch with user-design's values and therefore useful in other ways but not for the purposes of advancing user-design practice for instructional design. In order to help make sense of this litmus test, Table 3 lists the criteria which are both essential and necessary for a tool to work within a user-design framework. Please note that this does not mean that these criteria are in any way exhaustive; indeed, it would be impossible to reflect all the criteria necessary for any design process.

As Table 3 shows, some of these methods, on their face, or in the basic research/theories that have been put forth in the literature to this point, may not fit all the criteria I hold for a successful user-design tool. However, it is important to have the cognitive flexibility here to be able to adapt a tool which meets most of the criteria above and which you suspect may be useful in UD practice. It is easy to see how some tools simply do not fit and would be rather like square pegs in round holes. An example here may help to illustrate the point. The construct of "Participatory Rural Appraisal" (PRA; Kane, 1997) is on the face of it a strong possibility for user-design tool adoption. Closely related in practice to Participatory Action Research, the idea behind PRA is a deep and significant engagement with stakeholders in rural contexts in which stakeholders "influence and share control over development initiatives, decisions and resources which affect them" (p. 1). This is certainly a great start toward user-design; a power shift seems inherent in PRA. However, upon closer examination, it becomes clear that the theoretical underpinnings (which certainly will have sway on the practice) are too oriented toward generalization despite the recognition of context as paramount. PRA is focused on positivism and empiricism in ways that create a tension between its practice and its theories. According to Kapoor (2002), PRA needs more attention to problematizing and policiticizing its underlying theoretical constructs and clearly, the power shifts that are hoped for in user-design are incompatible with the underpinnings of PRA. Therefore, it is not really an acceptable tool for user-design; it is one that we examine, turn around, read over and over, look at the pros and cons, and then finally put in another pile of useful tools that are incompatible with user-design.

Thus, not all tools are possible user-design tools, but some that are near misses can be adapted. Such is the case with action-based user-design. In this case, the fundamentals of the parent tool (action research) is not as consonant with user-design as it might be, depending on which theorist you select as your foundations for action research. But, if we bend the construct a bit, it can be a fantastic fit. In the same way, design based research is not generally understood to be about empowerment (that is, not central to its tenets at the moment); however, we can adapt it slightly because it has many of the same foundational beliefs as user-design. The reason I am spending so much time explaining tool selection is to empower you, as an instructional designer, to identify and adapt new tools as they emerge in our literature for the purposes of creating new UD tools that work best with your own practice. The primary determinant, for me, is the underlying theoretical constructs and the way that they handle issues of power and expertism. Now that we know a bit more about how to select tools for user-design, let's look at these five tools and how they can be used in instructional design.

Each method is discussed in detail after a brief fictional scenario illustrates what would happen in the field if a given method of user-design were employed.

Ethnography

Ethnographic field methods, which are considered in our first scenario, generally involve an openness to the native or *emic* point of view (Vidich & Lyman, 1994) and extended periods of fieldwork most commonly in the form of observations, interviews, group interactions, and document analysis (Atkinson & Hammersley, 1994). Guidelines for research-oriented and critical ethnographic studies which focus on cultural understandings are readily available (among the best is Carspecken,

1996). Taking these basic ideas and applying them to the activity of design within a user framework is the focus of this first scenario:

SCENARIO

Lyssa owns a small catering business that she started seven years ago out of her home. The company hit it big about two years ago when they started offering small gourmet dinners to busy working families. But now the business has grown too large, too fast and Lyssa is concerned about the organizational culture of her group. There isn't a specific problem—there is not a chilly climate for workers or a higher degree of complaints from customers—just a sense that things are not quite right. Being a proactive business owner, Lyssa has decided to undertake an open-ended project to help her and her workers better understand their work and what, if anything, might improve their situation.

Lyssa calls her friend Austin, a consultant in ethnographic UD. She thinks it's a pretty fancy name for watching folks and talking to them ... rather new age in fact, but she knows Austin and trusts him. She's talked to him over coffee and dinner about what ethnographic user-design might do for her organization and so she introduces Austin to her workers in a brief meeting, explaining that Austin will be talking to them and watching them work. She assures her people that they are not being scrutinized for efficiency evaluations or raises or the like. Fred, an old-time chef, asks why Austin doesn't just conduct surveys or something? Austin explains that ethnographic user-design does not assume any particular biases or agenda since they are not really trying to solve a particular problem. Instead, Austin explains, this process helps the users and leaders understand their environment, their setting, their work—and from that understanding new systems might be built in collaboration with all the workers.

Austin proceeds by focusing initially only on events in an effort to describe the variety of happenings that occur everyday in the natural work setting (e.g., cooking, delivery, billing, etc.). He examines these without judgment and with an eye toward capturing the users' experience of these events. He conducts interviews, does many hours of observations and carefully examines workspaces. Eventually he finds that he is not seeing or hearing anything new—he has reached a saturation point and he turns his focus to the places where the work happens: the office, customers' homes, party halls. He conducts the same process in these places He also examines particular objects and finally focuses on the people themselves.

He compiles all of this information (e.g., observations and interview transcripts) and looks for emergent themes such as the frenetic pace with which all work is done as reported by the people, and as observed. Austin then meets with Lyssa and those he has observed or interviewed to describe the sorts of themes he has identified during the ethnographic contextual inquiry. Based on this information, the group decides they would like to carefully consider issues of worker stress and ways that would best help to alleviate this. This is not Austin's recommendation, it is what the users have decided to focus on based on his open-ended work.

Naturally Austin could have biased the work by creating a problem of stress that wasn't really there ... but the users would have rejected that notion because they would see they are not feeling stressed. And since Austin has not presented himself as an expert, the users feel free to correct his assessments. Austin has nothing to gain by suggesting stress is an issue, and since his work is purely descriptive and not prescriptive, it is up to the users to decide if anything needs to be done and what. At this point, Austin suggests a few resource books on stress management and a professional association which may be helpful in identifying stress-reducing workshops. Austin also explains the process that he used and outlines a modification of this process which would allow the users themselves to continually learn from similar organizational observations. Finally, feeling that the organization is equipped to continually understand and improve their culture, Austin removes himself from the setting.

Blomberg, Giacomi, Mosher, & Swenton-Wall (1993) have outlined the orientation of ethnographic field methods as the intersection of anthropology and design. The meshing of these disparate fields is not an easy task, but it is a systemic approach to both. Ethnographic participatory design (Blomberg et al., 1993) requires a period of field work where design has traditionally been done by designers alone or in design teams without prolonged exposure to the work space.

Ethnographic field work proceeds under certain central guiding principles, including studying performance in everyday natural *settings*, within the everyday natural *contexts* in which performance occurs. The field work is aimed at descriptive rather than prescriptive understandings—that is, understanding how performance actually occurs rather than how it *ought* to occur—obligating an ethnographer to take a nonjudgmental view of the performance being studied. The field work requires an openness to qualitative study (LeCompte & Preissle, 1993). Finally, ethnography is guided by a desire to interpret and give meaning to the insider's view of the work situation. Observations (Adler & Adler, 1994) and interviews (Fontana & Frey, 1994) are the primary methods utilized by those interested in ethnographic user-design.

Observation under ethnographic methods should remain unobtrusive with minimal disruption to work processes. Ethnographers distinguish between ideal and manifest behavior and note that most respondents to survey instruments or out-of-context interviews will cite ideal behavior—that behavior which is desired or preferred—as opposed to manifest or actual behavior, which can only be observed in the work context (Blomberg et al., 1993). Because it is not possible to observe everything happening in a busy work environment, ethnographers focus their observations by events, people, places, or objects. These observations are very important to the effective application of participatory or used-design since the emic or native view is continually sought to inform the design process.

Unstructured, open-ended, conversational interviews can be combined with observations and are typically a part of any ethnographic method. Since the social situation created by interviewing is artificial, the setting for the interview must be thoughtfully selected. In addition, comfortable settings enhance the ability of respondents to shape interactions during interviews about their performance. Remember you are doing *design* here, not research. Therefore, consistency in settings is less imperative. This is also true of design efforts; settings should not be intimidating or distancing. Instead, establishing rapport and recognizing that the user knows more than the interviewer is important in interviewing as well as in design activities. This recognition of user expertise moves the designer out of the expert role, which can be a disorienting experience. Blomberg et al. (1993) wrote of this particular situation:

> A related problem when ethnography is part of a design project is the tendency of some designers, whose job after all is to solve problems, to come up with solutions to a respondent's expressed problem before taking the time to get an adequate understanding of the problem. The push to solve design problems may conflict with taking the time to fully appreciate the nature of the problem. (p. 137)

Connecting ethnography and design in a way that designers can work alongside users is perhaps best done in balanced teams of ethnographers, designers, and users.

Cooperative Design

Cooperative design (Bodker, Gronbaek & Kyng, 1993) directly engages the users in the creation of their own systems of learning, software interfaces, or information systems. It recognizes the inherently political nature of creation and design and encourages the use of familiar tools to create used-designs):

> *SCENARIO*
>
> *Midwest Electrical and Mechanicals services all major mechanical and electrical systems in private homes and small apartment buildings. Midwest has decided to implement a new technology plan. The leadership has already bought into the importance and financial benefits of computerizing many of their functions, including administrative support, data entry and documentation via computers for house calls. Because Midwest is located in a small town and they have an excellent reputation, their repeat business is extremely high. They seek to personalize their services more by having running records of any given home, much like a doctor has a chart with each patient's medical history. In the past, they have kept track of such information through pencil-and-paper forms that are filled out whenever service is rendered to a homeowner or landlord. The paper is then returned to the central office where it is entered into the computer and saved for future reference. When the homeowner calls again for service, the individual record is printed up and given to the service technician before s/he leaves central office.*

Tom has recently been promoted to the Chief Information Officer position for Midwest. His task is to help with the training, implementation, and adoption process for this new computer format. He is in charge of the training as well as the purchase and evaluation of the new program. In a traditional model, Tom would look the situation over, do a needs assessment and begin designing training for the end-users without substantial consultation with them in terms of what they want. He would determine based on some observations and document analysis what the interface should look like and the type and number of machines to order. He would present this completed proposal to leadership and negotiate approval.

But Tom has recently read some work on User-Design and decides he wants to use a different approach because his past experience with innovation at Midwest has shown a large degree of resistance among users, particularly to expertise and top-down change processes. Past efforts have been particularly contentious between union and management and underlying this conflict has been a sense of working class disenfranchisement. Selling his leadership on the benefits of stronger ownership and ease of adoption, he sets up a team of users and leaders whom he asks to help him make these decisions.

The team begins by conducting a future workshop in which the problem of a new information system is presented to the 17 users and 2 leaders. During the workshop, users, primarily, are asked to critique the current situation, fantasize about what they would see as an ideal situation and then create a plan for how to get to that ideal. They are asked to do this with regard to the interface, the training sessions and the purchases. Leadership is asked in particular to help with the planning phase by finding concrete ways to facilitate the user suggestions whenever possible.

Tom works with the users to specify what it is they want in the interface and what would make for easy use. Eventually he creates cardboard mock-ups of these ideas and runs them by the user-design team, which makes a number of suggestions for improvement. Consensus appears not to be a problem as most of the users seem to conceptualize their work similarly and therefore want the same things in their computer system. As the project nears implementation, Tom uses cooperative prototyping to gain nearer and nearer approximations of the final interface. Users are asked to critique the still rough prototypes, which are no longer on cardboard, but rendered on computer screens, and then are asked to play with the "near" prototypes reporting what problems are likely to occur in the field. Tom then shares the proposed interface more broadly with the users to try to work out any final bugs and then builds the program completely. A similar process is used to design the training program and the purchase plan. Finally, Tom negotiates the final plan with a small group of representatives of leadership and usership, recognizing the inherently political nature of design itself.

Based on experiences with Scandanavian computer interface design, Bodker, Gronbaek and Kyng (1993) describe a variety of methods for engaging users in cooperative design activities such as future workshops, mock-up design, and cooperative prototyping. The work of cooperative design recognizes the inherently contested nature of design and rather than resting power in the hands of designers and/or leaders alone, shares it with everyday users by engaging them in design processes. Of course, this requires a demystification of design and friendly design tools such as the cardboard mock-ups mentioned in the above scenario.

The guiding principles underlying cooperative design include the political nature of design, authentic participation from users, a focus on the "use situation" (that is, how computers, or other innovations are actually used), working with design tools that are familiar enough for nonexperts to use, and simulation of future work situations (Bodker, Gronbaek and Kyng, 1993).

Three methods for engaging users in cooperative design include future workshops, mock-up designs, and cooperative prototyping. In future workshops, a common problem situation is shared among approximately 20 participants representing major user groups. According to Bodker, Gronbaek and Kyng (1993), such workshops are comprised of three main phases: the Critique, the Fantasy and the Implementation. Mock-up designs are cardboard creations of the actual technology to be considered (such as a new computer interface for data entry tasks). These rough looking approximations can help engage users precisely because they have an unfinished appearance that encourages alterations. Cooperative prototyping presents a nearer approximation of the final design but still creates a familiarity with design tools which empowers users to make changes to the design. Cooperative prototyping is similar to rapid prototyping (Tripp & Bichelmeyer, 1990) except that, again, the power remains in the hands of designers in rapid prototyping despite the opportunity to get more input from users. Design in cooperative prototyping is meant to be a shared enterprise. To the extent that the tools of design are themselves easy for users to employ, users can experiment with design changes in cooperation with designers. While cooperative prototyping is more expensive due to increased development time to produce more polished, finished prototypes, and the tools are not yet completely familiar to the everyday user, it does force a confrontation with the impact of the innovation in the work setting.

Action Research-Based User-Design

The third tool we examine is action research-based user-design. This tool asks communities of users to seek deeper understandings of their situations thereby building systems which speak to their needs. It is a broad enterprise which is fundamentally aligned with critical approaches and seeks to change power structures through the education of users.

SCENARIO

Cityscape schools has recently been taken over by the state legislature. Citing fiscal mismanagement, failing infrastructure, and poor student outcomes, the state has decided to suspend all normal working procedures in the school district, firing most administrators, the superintendent and all school board members. The state decrees that the district will have five years from the date of takeover to reorganize itself and design a new, more effective system. If, in that time, the community is able to present design documents to support their efforts, they will be given a chance to actualize a new school system using state monies again. If not, the state will remain in control of the district.

The community is understandably upset, but after some period of public hearing, they come to understand that they have an opportunity to improve their schools through their own work. Stakeholder groups gather to decide what to do next. About six months into the takeover, the community partners with a university and find a group of students enrolled in an action research course who are interested in helping them to understand their current situation better and solve their takeover problem. The students and faculty member negotiate an agreement detailing what they will do for and with the community. Their goal is to help the community understand the role of the school in their town. They observe classes, meetings, and interactions. They interview stakeholders, and examine curriculum and facilities. They interview many community members and begin to create charts and descriptions indicating the relationship of the school to the community. They collaboratively construct an understanding of the school inside and out.

About this time, the course finishes and the action researchers, believing that their job as researchers is complete, begin to extract themselves from the school district. But the community, realizing that they need to produce something beyond this mere understanding—something called a "design document"—renegotiate their arrangement with the researchers.

Based on the initial findings of the research team, groups of stakeholders and leaders meet and begin the difficult, at times painful, negotiation of a new school. Only one year into the process, there is plenty of time to create not only a design process that works for the community members, but also the design of an optimal learning system for this particular school–community context.

The community works through their understandings of change as an event—an end-point and instead create a continuous process for design, improvement, and reflection. The community's design efforts are focused on ideal systems of human learning, excellent curricular approaches, high standards, and helping all children become life-long learners.

Based on these ideals, community members (designers) found commonality in the desire to engage in year-round schooling, block scheduling, alternative assessment, core course curriculum, arts and music curriculum, special needs education and gifted education. Each of these initiatives were further designed in small groups and ideas were shared across groups.

As each group worked on their "part" they found it difficult to focus on a single issue and began designing systems that met their increasingly specific group desires.

> *Negotiation between the groups became less important as the district adopted a schools-within-schools model. At last all the designs came together under the umbrella values stated at the district level. A number of different schools were created to meet the needs of specific community groups and design efforts.*
>
> *The design documents were provided to the state eighteen months prior to the deadline. While all the specifications were not completely fleshed out, most of the critical elements were developed to the state's satisfaction.*

The merit of engaging whole communities in the understanding of their problems in order to create change systems that continuously seek to answer specific community needs is fundamental to current conceptions of action research (Park, Brydon-Miller, Hall & Jackson, 1993). Cases such as communities banding together to cause the government to assist in the moving of entire communities when dams are built (Comstock & Fox, 1993) or of the reclamation of Appalachian land (Horton, 1993) exemplify the values of action research which empower people to take effective action toward improving their lives and community structures. Founded on a deep respect for indigenous knowledge, the field of action research shares assumptions and underlying values with the practice of systemic change in education, and user-design.

Action research, like many recent qualitative method advances, actually represents a number of approaches to research (Reason, 1994). Anything from "teacher-as-researcher" to engaging entire communities in substantive change from a critical standpoint may qualify as action research. There are many approaches and methods perhaps most strongly associated with participatory or collaborative research (Torbert, 1981). The main thrust of action research is engaging people in understanding their situation so that they may act to solve a problem (Patton, 1990). The role of the researcher is facilitative, not expert oriented (Stringer, 1996), which is consistent with the fundamental values of user-design and systemic change.

As I use action research-based user-design, it is slightly different from the current conceptions of action research because of the design component. Action research-based user-design focuses on the combination of research and understanding of indigenous knowledge *with* design efforts to bring about change through particular action. Thus, engaging teachers in understanding their practices would also require a scaffolding of the design process in order to be considered action research-based user-design. It might be action research, but the design component is not present and this is perhaps the most critical distinguishing feature of action research-based user-design. Finally, action research-based UD is focused on self-reflection among practitioners or users (Argyris, Putnam, & Smith, 1985; Schon, 1983). While reflection is an important component of action research, tying this reflection to the process of design transforms reflections into ideal design images.

The major distinction of this method from the previous two methods is the unique focus on research and problem solution. Ethnographic user-design focuses on a deep understanding of the native voice and the creation of artifacts which support that voice without specific identification of a problem situation. Action research is more focused on responding to problems through idealized design activities. At the same time, cooperative design methods are not focused on inquiry or research in any specific way; only as an incidental outcome would research understandings occur in cooperative design efforts.

Design-Based Research User-Design

Design-based research is an outgrowth of the learning sciences movement which focuses on the examination and furthering of design theories through primarily experimental research. This flavor of research has gone by many other names in the past such as formative research Roma, (1990). Reeves (2005) points to a variety of different terms that encompass design-based research including "development research, design research, design experiments, design studies, formative research, formative experimentation, and engineering research" (p. 50). However, it is not necessarily the case that design based research is more "friendly" to experimental or more expert-driven models. Much the opposite, design based research theorists recognize that this method is not meant for generalizable findings. Design based research proceeds through a series of stages, steps, and phases based on several principles that are explained below. While design based research is not necessarily meant to be used for the purposes of effecting change in practice, or to change design as a process, it is clearly aligned with user-design in their engagement of users and concern for learner centered orientations:

> *SCENARIO*
>
> *The Red Cross of America has explored many ways to effectively reach out to communities in need. Having paid good attention to the needs of those affected by recent floods in America, the Red Cross is currently concerned with the serious needs in Somalia that are emerging and creating a significant human rights crisis. Samuel has been charged with training the team that will go to Somalia in response to the crisis Samuel begins his work by examining similar past projects to derive what contextual and process factors are essential to their work in Somalia and they create a basic theory of how to approach the crisis of humanity brewing in Africa. Compiling several white papers and field reports from past efforts, Samuel puts*

together a plan based on the theories generated from earlier works and that plan then becomes a basic design for the training. Samuel also consults the latest literature in learning theories and is particularly influenced by learner-centered theories and constructivist design specificiations.

However, before Samuel sets foot in Somalia, he understands a strong need to engage the users in their own design of training and decides that he should discuss his design with the eight other workers that are going to be heading to Somalia as the early response team to begin the work. He knows that this team will then bring back learnings to be shared with the larger group of Red Cross responders and that everyone is very busy and anxious to get to Somalia soon. Delays may cost lives and so time is of the essence in this case.

Having defined the basic design, Samuel modifies the design along with the eight other trainees as they engage in a session to determine what they most need from the training, what additional information needs to be sought, and what specific steps need to be taken first to begin the process of deploying to Somalia. The training design is analyzed through a variety of lenses by all users—through the typical cognitive lens, but also examining necessary resources, interpersonal requirements, and contextual factors such as classroom, group, and institutional/organizational needs. After these initial modifications and analyses by the users, the training is conducted for the initial group. And the eight initial responders depart for Somalia. While additional time had to be taken to modify the design given user-input, and to analyze the design from a variety of perspectives, that additional time seems to have paid off because the training went smoothly and upon deployment, the Somalian team has great success in initial response to the humanitarian crisis.

Naturally there were lots of things that this first team learned as a result of their work in Somalia and they recognize that the second wave of training can be improved based on their experiences. While on the airplane home, the team of eight begin to revise the training design that they had experienced in preparation for the second wave training. However, Samuel is not part of that airplane conversation, and when he finally joins the group and hears about the new training program that they have designed, he is hesitant. Realizing that the first response team feels a sense of expertism about the situation the second wave team will be facing, Samuel wants to hear the team out. However, he also knows that the power of the training design was largely in the user-design approach that was employed. Therefore, he asks for a modified approach, taking all the good ideas, and asking both first and second response teams to work on the training together. The first team is distraught by this suggestion. While they recognize that their own user-designed training was an excellent experience, they fear that engaging large groups in that process will be time consuming and cost lives. Samuel suggests a systematic and purposeful approach that includes several research methods such as observations, interviews, surveys, and document analysis. The observational data is primarily that which has already been collected during the time in Somalia, and the documents are a combination of Somalian documents and those that were used for the initial training sessions. Interviews are conducted with the second wave trainees and all of these processes are engaged in an ongoing way that continually informs the design of the next wave of training. The team agrees, while they are concerned about the additional *time that will be required, they also realize that this iterative approach will produce the very best training and that since it is iterative, there is not really a beginning and end and so the process can be seen as less pressured. Following the basic principles of design-based research within the framework of user-design, the training for first and second wave Red Cross workers create an extremely strong and effective training session which has excellent buy-in from multiple stakeholder and learners.*

The principles of design-based research as laid out by the Design-Based Research Collective (2003) are followed in this example. The design is founded on research from the start and practical goals are set from the start. The initial plan has to meet certain time constraints and those are attended to. The design is then taken to the users prior to implementation and conduct. This is the major departure from design-based research practice and principles. Once the design has been modified via an initial round of UD, the team implements the training and they collect research data on the results as they implement their humanitarian efforts in the field. During this stage, of course, they closely collaborate with participants. As a result of this first round, the users realize that they must implement systematic and practical research methods, examine that data and help it to inform their future training efforts while still including all users/learners in the process of creating their own learning experiences. Some of the principles of design based research are not utilized in the example here; for instance, validating the generalizability of the design is not an appropriate goal for a UD-oriented application of design-based research.

Design based research understands participants not as subjects, as they are understood in traditional psychological experimentation, but rather as participants who hold certain expertise that may help in the production of an improved design. The shared nature, embracing multiple perspectives and socially negotiated meanings, of design-based research is certainly appropriate for the user-designer interested in improving designs. However, the tendency toward experimental language including consideration of variables and outcomes means that traditional design-based research (Barab & Squire, 2004; Collins, 1999) is not a precise match for the user designer interested in engaging design-based research as a tool. However, design based research is interested in flexible, iterative, field-based research which can certainly be adapted easily to work within the user-design instructional design framework. As Hoadley (2005) writes, "Design based research boils down to trying to understand the world by trying to change it" (p. 46) and in this way, design based research is clearly aligned with the interests of user-design in both design and change; and in order to best achieve these goals, one must also come to deeper understandings. Therefore, design-based research has many connections to user-design and can be used in constructive ways to further user-design as a tool.

Collins, Joseph, & Bielaczyc (2004) set out a series of basic steps for carrying out design based research including Implementing a design, Modifying the design, Analyzing the design from multiple perspectives, Measuring dependent and independent variables, and Reporting on design research (p. 33). This set of steps does not seem all that different at the macro level from traditional experimental research design other than the clear focus on the design as the subject of research. Modifying it to more fully embrace the user, forego generalizability in favor of contextualization, and allow for significant power shifts makes design based research a useful construct for iterative user-design.

Scenario-Based User-Design

Scenario-based user-design is a modification on two streams of work; scenario planning (Schoemaker, 1995) and scenario design within the field of HCI (Carroll & Rosson, 1992). The wonderful thing about scenarios used in either planning or design is that they are extremely future oriented. And as we examine other similar movements such as *Preferred Futuring* (Lippett, 1998), we see that there is a significant power in scenarios which generate rich and vivid images of potential futures which, in turn, generate significant discussion and conversation among users who have something more concrete to relate to in the processes of user-design. Scenario planning is not intended for user-design and, in particular, it tends to maintain a high level of power and expertise in the hands of management and established decision makers. Nevertheless, the concepts are extremely useful for the purposes of user-design. If recast as an empowering tool for broad engagement in user-design, the use of scenarios in the macro sense of scenario planning (as opposed to the more micro sense of scenario design in HCI) is perhaps one of the most powerful tools introduced in this chapter:

> SCENARIO
>
> *The University of Great Lakes, a large Midwestern university with more than 40,000 students across 12 campuses in three states, has struggled in the last seven years with the overwhelming nature of technological change that has been occurring. Students arrive at campus with ever-increasing expectations for wireless connections, online course availability, computer lab facilities, and systems that are entirely automated and available online for everything from parking passes to grade postings. Other factions from staff to faculty to parents have equally high expectations. This pressure has caused the President to seek broad engagement with all*

stakeholder groups in the university in the form of user-design. The President is interested in seeing potential directions that the entire university might move in terms of technological change and ways to train new students, faculty, staff, parents on the dynamic technological systems in an ongoing fashion. At the moment, the university responds or reacts to each new change as it emerges. When a new course management system arrives on campus, after some sub-committee of technology specialists approve it, the training program is typically turned over to the development group and usually hastily constructed for all audiences to get them rapidly up to speed on how to use the new system(s).

The President is interested in being more proactive and anticipatory. He has learned about scenario user-design and decides this would be potentially scalable for the purposes of engaging large groups of system users and helping to set a course for technological futures rather than reacting to each new development.

Mr. President decides that it is not appropriate for him alone to engage in the processes leading up to broad engagement with the users and so he establishes a futuring committee who helps him to define the scope, major stakeholders, basic trends, and key uncertainties. These are the primary inputs to the creation of potential scenarios. The futuring committee expands out to a larger group of stakeholders and invites them into the creation of scenarios. The larger team, by examining the scope, stakeholder groups, trends, and uncertainties, create a set of initial scenario themes and then discuss them to ensure consistency and plausibility in the themes. These eventually move toward more specific future oriented scenarios and four divergent scenarios are generated by the stakeholder team.

These scenarios are then taken to the larger user group, an open invitation is issued through the school collegiate newspaper and list serves and publicized through the public collegiate radio station inviting as broad a group of users as possible to react to the scenarios and help make sense of how the training for these potential scenarios could most effectively evolve. 12 large group meetings are held in which the scenarios are presented to the large groups and technology is used to help the users offer feedback to various potential futures, preferred futures, potential training solutions associated with those futures, and finally to volunteer for further work to refine the technological future of Great Lakes.

The process described in this scenario follows the basic outline offered by Paul Schoemaker (1995) in his description of Scenario Planning for strategic thinking. It is vital to recall that strategic planning or decision making is quite different from the process of design and creation. However, the basic outline of steps offered by Schoemaker for the use of scenarios can be easily adapted to the processes of user-design. One of the important criticisms of user-design is that it may not be sufficiently scalable. And it is the case, in general, that user engaged processes are less efficient in terms of design processes, though they are highly efficient in terms of adoption rates and implementation processes.

Schoemaker (1995) offers three classes of knowledge that are important to consider in scenario planning: "1. Things we know we know;

TOOLS FOR USER-DESIGN 41

2. Things we know we don't know; 3. Things we don't know we don't know" (p. 38) One of the powerful sources of mitigating the classes of knowledge that are incomplete is the process of user-design. By engaging users, we can begin to see our blind spots more clearly and thus enlighten some of the uncertainties inherent in any futuring activity and particularly prevalent in scenario planning processes.

It is important when discussing the power of scenario planning to point out that even the earliest scenario planning advocates (Porter, 1985) do not suggest that scenarios are forecasts (That is, we are not seeking here to try to predict what will happen, but rather offer several potential images of things that might happen which can help us to be proactive in the creation of support systems of all sorts as well as giving a certain amoung of power to partially affect the future). Chermack, Lynham, & Ruona (2001) offers an excellent review of the literature surrounding traditional scenario planning as a management and decision-making tool. They identify a number of paradigms that guide scenario planning as a field and a set of methods such as reference scenarios, procedural scenarios, industry scenarios, and soft creative methods approach. Van der Heijden (1997) offers several criteria for scenarios including a minimum of two plausible scenarios (necessary to identify uncertainties), scenarios must be internally consistent and highly relevant (in our case, to user concerns), and scenarios need to produce a certain amount of creativity—new and original perceptions (Chermack, Lynham & Ruona, p. 16).

CONCLUSION

When approaching a tool box and determining which tool will be the proper one to use, any good craftsman carefully examines the problem and will consider the context in which the problem is set. It's also not unusual to try one and then try another or use several tools to fix a given problem. This is the sort of flexibility that is needed when engaging user-design as well. The first step is determining whether or not the problem really fits with *any* user-design tool. user-design is best engaged when the leadership is able to let go of control, the community shows some capacity for active design activities, the problem is systemic and expertise is not likely to help the situation or may be rejected by the community. The second step is a careful examination of the context; a global perspective on the problem. The craftsman doesn't try to repair a leaky water pipe without first determining how much water is leaking, where the leak is located, what is around the pipe—can I get to it with this sort of wrench, or this one?; what is the flooring beneath the water leak, where is the drywall in relation to the leak, etc. It is therefore imperative that we, as UD craftspeople carefully examine all of the contextual factors contributing to the problem that we hope to employ UD to rectify. The third step is determining which of the many tools make sense to engage user-design in a given context. Remember that you may find the need to create your own tool, modify the tools given, combine tools, etc, but that, they should align relatively well ideologically with user-design.

This chapter has presented several tools for user-design including ethnography, cooperative design, action research based UD, design-based research, and scenario-based UD. Guiding questions to help you decide which tool might work best for a given problem in a given context are offered in Table 4. Each of these methods can be extremely powerful and useful when employed faithfully, appropriately, and within the epistemological framework of user-design.

TABLE 4
User-Design Decision-Guiding Questions

Variable	Question
Appropriateness (for user design method) revisited	Is expertise useful in this situation? Would expertise be rejected as "not invented here" syndrome? Is the problem narrow or systemic? Does the community currently have or does it show capacity for active design participation? Will leadership support full user design?
Global perspective (questions to ask for all user design projects)	What interconnections and interdependencies are present in this system? Can I chart them? What does the system look like graphically? What does this global understanding of the system imply in terms of the ripple effect of change in this organization? In what larger context is the system of interest embedded? (What is the environment?) What is the idealized vision of the whole? What can I do as a designer to help negotiate the needs of users and leaders and to encourage user design? Is this a problem-based or design-oriented situation? (Indicates which user design method to use)

(Continued)

TABLE 4
(Continued)

Variable	Question
Ethnography	What natural settings do I want to observe?
	Who should I observe and interview?
	What are the important sources of design information?
	What would it mean to have an insider's view of this work?
	Do I want to use fieldnotes with or without videotape and additional aids?
	How can I locate, schedule, and structure my interviews in an open way to encourage open rapport and authentic engagement?
	How can I build a partnership with the users that allows me to be ignorant?
	Who are the end-users? Can I schedule time to observe them working? Are there secondary users I should observe and interview?
	Is the conversation during my contextual observations moving between reflection and engagement? How can I assure this?
	How can I expand my focus; what things have I been (deliberately or subconsciously) overlooking?
	How can we (as a team of inquirers) make sense of all the data we have collected?
Cooperative design	What are the power dynamics in this organization? Who has it, who doesn't and who who does the implementation of the innovation benefit?
	How can I encourage users to let go of rational decision making and really dream?
	Might I use future workshops, organization games, mock-up designs, or cooperative prototyping?
	What advantages or disadvantages do these techniques have in this context?
Action research-based user-design	How can I help this community better understand itself?
	What are the critical system elements that I might focus on for this work?
	What are the patterns of use, interaction, and communication that I can find, chart and diagram?
	How can I facilitate this community's learning about itself in a continuous way?

(Continued)

TABLE 4
(Continued)

Variable	Question
	What systems are the community suggesting we can create that will better meet their needs? How can I engage the community in the design of such a system?
Design-based research	Is there sufficient source of theoretical grounding to inform initial designs? Are there sufficient opportunities for iterative processes? Is there time and sufficient resources to support empirical data collection and analysis? Are multiple perspectives available to inform initial and subsequent designs? Are adequate resources available and prioiritized to report findings from the iterative design process?
Scenario-based user-design	Are uncertainties in future states of significant planning interest? Can users help to inform what leaders may not know or may not realize they do not know? Can at least two plausible, internally consistent scenarios be generated? Will futuring activities have real potential for impact on current practices? That is, is there sufficient power to change in accordance with the findings of scenario planning and designing processes? Is the problem best served by being more proactive than reactive?

Describe the relationship between user-design ideology and tool selection.

Which tool is most appealing to you personally? Why?

Which tools are most important for empowerment? Engagement? Powershifts?

Which tools are most useful for design? Holism? Change?

What are the critical steps to take when deciding on a UD tool?

4

Facilitating User-Design

INTRODUCTION

This chapter will begin to help you, as an instructional designer, to see your role in very different ways. We have always understood the purpose of instructional designers to be the primary resource of expert knowledge in basic educationism. But user-design turns that expertise on its head. It is vital to recognize that letting go of expertise does not mean losing your job or losing some part of your identity. Rather, we have to understand instructional designers' roles within user-design as less solution oriented, less expert, and more facilitative. In the same way that we see the teacher moving from an expert "sage on the stage" to a "guide on the side" in the information age, and the spiritual age to come,[2] the instructional designer's role will necessarily have to undergo radical transformations to fit in with the revolution inherent in the shifts from agricultural to industrial to information age learning. A discussion of the impact this will have on your role as an instructional designer, along with a set of specific steps you can take to facilitate user-design to assist with effective utilization of these powerful tools, concludes the chapter.

Letting go of the old is difficult letting go of what makes us feel secure in the world, secure in our work lives, in particular, is almost impossible. It requires a significant shift in people's identities and this is precisely

---

[2] I believe that we have seen several waves of societal change and that the next wave will be one of spiritual foundations. Evidence of this is already on the horizon with the political sphere being more influenced by "values" than ever before and wars being fought in the name of "religion."

what we've been asking teachers to accomplish for years in order to effectively employ constructivism or powerful uses of technologies in their classrooms. User-design asks the instructional designer to confront just such a sacrifice just such a dramatic identity shift in our own work world. First we need to identify the power dynamics inherent in traditional ID and contrast that with user-design. Then, in order to fully understand how user-design impacts the role of the instructional designer, we must explore critical systems theory in the understanding of changing power dynamics, empowering users in the design process, and implications for the practice of instructional design.

User-design is crucial because, as the late Dr. Bela Banathy has told us, it is immoral to design a system *FOR* someone else rather, it is only appropriate and ethical to engage users in the design of their *OWN* system. Note this term, *immoral,* in an age when values are becoming increasingly important, and I would argue, as we sit at the precipice of a new age, no longer the information age, but now the spiritual age, recognizing what is moral and immoral is paramount to ethical practice in any field—instructional design included. While you may or may not agree with the ethics of designing for others, it is still difficult to let go of one's identity as an expert in ID in order to let others "in" to that process in order to share the power of knowledge and technique with users.

While the ethical rationale for engaging user-design methods is certainly compelling, there are other reasons for such empowering approaches. Perhaps one of the most compelling is that those products and processes which are designed by users will build ownership among users and create a very different adoption cycle than we have typically seen through more manipulative and colonial (Rogers, 1995) models of change and adoption. Thus, truly engaging users in the creation of their own systems is not only ethical, it is also practical.

However, there is a huge diversity of stakeholders in almost all of our systems. And thinking about systems design on a grand scale, something as large as the design of a new school, may seem overwhelming or impossible. Imagine trying to bring the community of stakeholder interests together in the effort of creating a newly designed learning system. "There are too many differences of opinion, we cannot agree on anything," may be the cry of the novice user-design facilitator. This feeling is only slightly lessened when dealing with more constrained learning problems such as training firefighters for emergency procedures or training teachers how to use a new video camera.

It is true that the diversity of stakeholder needs cannot be underestimated as an obstacle to user-design. Perhaps the most effective way to deal with the diversity of stakeholder needs is to face, head-on, the limited ability of any organization to meet all the needs of stakeholder groups rather than promising to meet all needs and subsequently being unable to. Here is where, too often, we fail. We pay lip service to the voice of the stakeholder by including them on a team or a council charged with giving "input." But, that input is rarely communicated directly to designers, and even more rarely heeded by the designers or leaders. And it is naïve to suggest that simply putting all the stakeholders into a room (if you can get them all into a room together) can cause significant change to happen, can create a powerful training design, or can overcome hardened and established rules of power already inherent in every organizational context. We must recognize that UD is a very difficult enterprise and that promising people power without delivering is worse than doing nothing at all. This is because inauthentic participation has led to a great deal of cynicism among many stakeholder communities. Instead of this lip service, perhaps we need to show stakeholders that we take their needs seriously and alongside them design flexible systems that can meet their needs and in which they recognize the limitations because they helped to design the thing, afterall.

POWER IN USER-DESIGN

Engaging user-design and consequent role shifts means a significant change in organizational power dynamics. In traditional design, and, I would argue, even in user-centered design, the tools of design itself are the realm, even the possession, of the designer. These strategies and processes are assigned a complete set of rules and jargon-filled terms to describe the processes of design. Designers then study these terms, which many consider largely common sense with a good deal more language wrapped around it, and become "experts" in design— ID or educational systems design (ESD). Armed with this expertise and language and usually some form of certification such as a masters or doctoral degree as identified by a university diploma, designers guard against sharing the knowledge of design processes too broadly lest others become aware that they can "do" design pretty well themselves. I realize what I'm saying here is tanamount to "the emperor has no clothes" Let us recall who said the emperor had no clothes it was a small child, a member of the community with the very least amount of power. The child cannot see everything that the adults can see. And indeed, design itself is a legitimate intellectual domain, and if left entirely to their own devices, users may miss important design opportunities—just at the little child may miss some of the things that the adults who have a different physical perspective on the parade going by can see. Yet I state the issue this starkly to make the point that the tools of design are not so impossible for users to be able to understand and employ themselves. After all, the little child could see that the emperor had no clothes, while some adults were blind to it, some embarrassed by it, and some simply too embedded in established protocols to admit it. While I see the process of design as a true partnership, stating the question in such almost confrontational language forces us to face the issues of power that result from our control of the language and technologies of user-design.

My approach to refocusing your skills as an instructional designer centers on issues of power. user-design represents a radical departure from current power dynamics in instructional design because it not only puts user needs at the center, but also asks users to engage in the design process itself; no longer reserving design decisions for the designers alone to forward to leaders for approval. This deeper engagement is perhaps the least efficient way to go about design; it requires a lot of communication, but it is perhaps the most efficient way to go about diffusion and adoption.

It is my view that the only way to significantly change any sociotechnical system is to move from a power structure which focuses on experts to one which involves all stakeholders. Visions of new school systems, new constructivist learning environments, new online high school programs, new training systems, reformed teacher training, and new instructional methods are all admirable but do not address the underlying changes necessary for fundamental change. The design process needs to be shared in a culture of open collaboration among users, learners, parents, community members, professional educators, instructional designers, trainers, and (when possible) children. If we bring to those collaborations a set of clear visions or ideas about what the final product should be, we are necessarily limiting the opportunity to encourage a truly open collaborative culture.

As an example, if we consider what community participation in public schools could be, we might imagine schools where parents, educators, administrators, and learners collaborate to create dramatic, customized changes in their schools. However, when we look at what community participation in public schools has been, we see an unfortunate sham in which parents are promised voice but are then given no legitimate power (Yaffe, 1994). Far too many community teams in recent reform efforts have been turned into "rubber stamp" committees whose sole purpose is to approve all that the professional educators offer them in order to give the appearance of sincere community participation and voice where there really is none. This reality can be understood through the sociotechincal lens of expertism. In general, experts and professional educators often believe they "know what's best" for kids—possibly because of their experience with research findings or possibly because of their own hands-on experiences watching kids and schools, but, the end result is a clear idea on the part of the educator which is not necessarily shared by the learners of the parents or the community at large.

The potency of stakeholder participation lies not in its ability to co-opt political support for decisions pre-fabricated by experts—be they teachers, administrators, researchers, professors or instructional designers. Instead, the power of participation comes from shifting power and responsibility to members of a community who are all invested in educational outcomes. This power may come at the expense of efficiency. It may be forced through the current system and structure and require inordinate amounts of communication from experts to non-experts. It may even be rejected by the users who may claim that this is not their job and that the demands of postmodern life are too time consuming to allow them engagement in any serious way in design processes. Users have very little power in the current system of design. And I hope that if given a sincere opportunity to participate, most users would indeed surpass apathy and cynicism and truly engage in design experiences. The key here is the affordance of a sincere opportunity—they must be given power. Without a power shift, blighted hope is all we can encourage by embarking on a used-design agenda. Participation invites ideological conflict among divergent factions which can be both a source of energy and a source of vexation. Chapter 5 will address conflict as a source of energy in more depth.

FACILITATING BEYOND APATHY

There are several important obstacles to the effective application of user-design in instructional situations. Among these are lack of leadership support, inadequate resources, and even impatience. But, perhaps the most common and stubborn of obstacles is apathy among users. Over the years, my work has continued to shed light on the reluctance of stakeholders, particularly disenfranchised stakeholders, to participate in design processes. Some of those I have interviewed (parents, users, workers, etc.) say they are reluctant to participate because they see these teams as highly unlikely to make any substantial difference. They say that they are willing to make sacrifices in their lives, but only if they feel that they'll have power not just a say, but if they feel that they can really make something happen if, as they claim is too often the case, the leaders back down from users' bold proposals. If we tell stakeholders that they will help to shape a change or design a new system, then we'd better be prepared to deliver on that promise. This will probably mean leaving our own visions behind and, instead, accepting that this is a negotiation process over which we should have no power or authority and instead a faciliative and contributory role.

The purpose of this discussion is to identify strategies for combating the apathy in our communities and for encouraging an effective design culture in our society. Design is venturing beyond the known to the unknown—beyond the planning to the chaos (Hlynka, 1996). It is not offering something already known in a new visionary package of uniquely interacting components. Indeed, such jumping out can really only happen when all the stakeholders are consulted and engaged in the sometimes painful process of negotiating shared visions and values.

Offering users substantial power in the creation of their own tools means a shift in power relationships. It is no longer appropriate for us as designers to somehow "own" certain knowledge and information about human learning. In the past, knowledge has been seen as power itself, but primarily in its capture and unequal distribution. Perhaps instead of offering visions that communities are likely to accept as the recommendations of experts, we need to empower them with their own visions. A commitment to user-design carries with it a powerful obligation to empower others.

As you might imagine from the above discussion of user-design, actually engaging users in understanding the tools necessary to authentically participate in the creation of their own systems of human learning takes more than just an open invitation. In many cases users feel completely beaten down, disempowered, and/or cynical about their organization, school, community, and even their nation. I would argue that apathy is truly borne of cynicism; a feeling that the one voice doesn't hold power, because those who do hold power don't really allow everyday people a serious voice in decision-making. This sort of political power dynamic has robbed us of our power rather than democratically empowering us. And voting is the system by which we claim we're being responsive and open, when, in reality, the game is "rigged." Those in power maintain power, and will earnestly resist any shifts of power, away from them.

In Pennsylvania, a classic example of this was the recent Act 72 (2006) legislation at the state level. The Governor of our state, Ed Rendell, proposed legislation which was enacted at the state level allowing revenues from gambling in Pennsylvania to be applied to property tax relief. However, participation in the program was optional and came with some strings. The school boards were the decision-making bodies for determining whether or not a district would participate. Governor Rendell believed that public pressure for lower taxes would force school boards to capitulate to the additional requirements of the legislation and that they would vote to participate in the program. One of the key strings attached to getting access to the gambling revenues was the agreement of the school board that they would not increase property taxes more than inflation in any given year without enacting a referendum to allow for a complete vote of the public on any increased taxation. This meant that school boards, who currently have the power to determine taxation in Pennsylvania for their district homeowners, were being asked to voluntarily relinquish this power by voting to participate in the optional program. Despite a number of letters to the editor, there was not a groundswell of pressure against boards who did not support Act 72 participation statewide. Only about 20% of all school districts agreed to participate in this first vote. They were told that they would not have the opportunity to opt in later, and that those who *DID* opt in would be given proportionally larger shares of the revenues. This is a classic example of power issues. Those in power are never going to willingly relinquish power. The school boards within the state, even with the promise of high tax relief for property owners, did not want to give up their power to pass budgets that increased property taxes beyond the rate of inflation. They may agree under pressure (a factor Rendell counted on but never manifested itself), but they are not likely to voluntarily give up that power. We can see legislation either as a set of constraints that we must, as facilitators, help user designers meet, or we can see them as a way that the powerful attempt to maintain power and fight against those legislative mandated changes.

Doing the work of design is hard. It requires a lot of hard thinking and debating, discussion, and conversation. No part of this process is easy. However, in the traditional constructivist notion of learning by doing, it is clear that the process of design offers great hope for not only increased adoption rates and ethical design practice, but also for human learning (Wiggins 1998). In the same way that students who have been enculturated into the learning processes of traditional classrooms complain about the work of active learning of any sort, we can expect similar reactions from any users who are asked to engage in the design of their own systems.

Here is where the critical systems theory comes into play. The intersection of critical theory and systems theory comprises critical systems theory (Flood & Romm, 1996) that recognizes the overlap in fundamental values underlying both social systems theory and critical theory while extending them both to consider traditions not typically part of the discourse in either field. Educational Systems Design also shares some of these fundamental values such as the importance of engaging users in the design of their own systems rather than manipulating them into accepting designs of expert origins. Social systems theories have always embraced the "softer" side of understanding wholes. Unlike their counterparts in hard systems theories, social systems theorists have fought long and arduous battles to forward the agenda of a valued systems theory—one that values humans placed in a post-cartesian context. At the same moment, critical theorists have been waging their own wars against the modernist interpretations of life as equal and fair and democratic for all. The main project of the critical theorists is the uncovering of the "contradictions, social inequalities, and dominances" (Nichols & Allen-Brown, 1996, p. 226) in which capitalism remains a significant issue. Educational critical theorists like Henry Giroux (1992), Mike Apple (1986, 1988, 1990), and Hank Bromley (1992) have been raising critical questions such as who benefits from a given policy innovation, who is hurt, and what are the economic implications for a variety of stakeholders? Critical theory attempts to examine an innovation, a policy, even a current cultural state from as many angles as possible—most frequently from unexplored angles such as those perspectives represented by the underpriviledged in society.

Critical theory is particularly well-suited to the user-design project for several reasons. First, action research is one of the cornerstone tools of critical theory and it is also the primary, perhaps the only, way to research user-design. Second, several Habermasian assertions are clear theoretical foundations for UD (Habermas, 1984, 1987, 1990) including the lack of coercion for rational participation of stakeholders, a holistically structured knowledge, and the rejection of "an elitist splitting off of expert cultures" (Habermas, 1987, p. 330). Finally, although I do not wish to conflate critical theory with postmodernism, postmodernism is concerned with power shifts and the "dethronement of the authority" (McLaren, 1994, p. 196). However, postmodern theory does not embrace a systemic perspective and, in fact, rejects systems approaches as metanarratives (single best answers). However, there is a side of systems theories which is being ignored by the critical theorists in the same fashion as the systems theorists are ignoring the values of critical theories. Thus, the meeting of these two manages to inform one another in important and synergistic ways in the critical systems theories of Flood & Romm (1996).

WHAT DOES ALL THIS MEAN TO THE PRACTICE OF INSTRUCTIONAL DESIGN?

Perhaps the most difficult thing about what I am proposing in this chapter is that it requires substantial shifts in the identity of instructional designers and the field of ID as a whole. If we are to open up the processes of ID to this critique of expertism and authority, of metanarratives and inauthenticity, it requires of us a confrontation with the limitations of ID. Clearly, the processes of traditional ID models (Dick & Carey, 1996; Seels & Glasgow, 1990) are metanarrative in nature, that is, they present one right way; they do not revel in the diversities of possible ways or present oppositional ways. Even the most promising of "progressive" instructional design models (e.g., user-centered (Sugar, 2000), appreciative ID (Norum, 2000), responsive ID (Ertmer, 2000) still advocate particular approaches with specific ways of proceeding—therefore, they too are vulnerable to the critique of a metanarrative, as is even this text.

Seeing user-design and the consequent shifts in power that may lead to significantly increased adoption rates and fidelity to the designs we create means leaving behind some of the science of instructional design. It means educating the user, to the extent that it is possible to do so without upsetting serious power dynamics (that is, we are not light bringers, but we might have some ideas that are worth sharing if users want new ideas in order to help them design their own learning). It means sharing decision making broadly even when that is not a highly scalable activity. It means letting go of some of our orthodoxies such as terminal objectives, test item analysis, and learning transfer and deconstructing them to allow users to directly access the ideas inherent within these constructs. So, user-design means that ID must undergo a radical change; a shift from a linear contained set of jargon filled steps to be taken by experts into an open, fluid, iterative process engaged in by whole communities. This is an uncomfortable answer to the question posed for this section. I cannot simply put down on paper a new model; I cannot lay out a new set of steps. Much the opposite, the result for a user-design practice of instructional design is a contextualized, non-scalable approach to creating new systems of human learning.

SOME STEPS FOR FACILITATORS TO TAKE

Despite this rather nebulous state we are now in, letting go of specifics and entering into the world of power shifts, there are a few things that the facilitator can do to help prepare for the practice of user-design ID. These steps are not the same thing as the steps or stages in the user-design process itself, but rather are some specific steps that facilitators can take in the process. A more complete description of the stages of user-design as they compare to the steps in the ID process is shared in chapter 9. These facilitator steps are in no particular order:

i. Let go of expert/light bringer expectations
ii. Carefully examine your own power issues
iii. Look at the context you are embedded in for power issues. Chart it if you can
iv. Anticipate user-reluctance and brainstorm some ways to overcome
v. Ensure leadership buy-in
vi. Select tool or tools, possibly in consultation with users
vii. Examine the system critically—who benefits, who loses, and use this information to further understand the power chart you created in iii

Each of these steps is worthy of brief explanation. Clearly in the first step the instructional design practitioner has to understand all of the material we've covered to this point about the colonlialization of traditional ID environments and to deeply examine their own abilities to let go of the expert power position. This leads easily into a consideration of the power issues that the instructional designer carries with them. It may seem "touchy feely" but it is essential that you know yourself well, that you rid yourself of self-doubt and insecurities that tend to exacerbate power sharing, and you know whether you are battling with the boss or the learners because you had a fight with your spouse or an old battle with your parents. A little psychology can go a long way in helping you become the best instructional designer you can.

It is also essential that all instructional designers understand the context they are embedded in and the power issues that are already present. This can take a good deal of time and can help to lead to some important answers to questions that are related to a typical needs assessment. If possible, it's extremely useful to chart the power dynamics and the context as a whole in some illustrative way that is meaningful to you and, hopefully, meaningful to those in the context as well. Knowing that many users will be reluctant for their own reasons, it is important for you to anticipate the typical objections raised and be prepared to answer them. For example, as we've seen in constructivist classrooms, active participation is often daunting to learners. In the same way, users may feel that they are paying you to bring expertise to bear on the problem, not to get them to solve their own problems.

So, be prepared to show users why it is important for them to be actively involved in their own design activities. Brainstorm ways to illustrate these important issues, role plays, exercises, even games can all help illustrate the value of user-design within your context, and these can help you overcome objections and reluctance on the part of stakeholders.

It is essential that leadership agree early on that user-design is the direction they want to pursue. If they change their minds along the way, or if the leadership itself changes hands, it is imperative that the facilitator carefully negotiate and renegotiate the user-design power of decision making (not just input). Without leadership buy-in for this process it is far more likely that you will meet with failure rather than success. In fact, the process may turn out to be a terrible error in terms of time investment and future user-confidence. Based on the tools outlined in chapter 3 of this text, the facilitator, hopefully in consultation with the users, should choose an approach and set of tools or single tool to use as a team in the design process. To the extent that it is possible to engage users in this decision, a certain circular user designed user-design process emerges.

Looking at the system from a critical standpoint understanding who wins and who loses in the current system is an important addition to the contextual chart that was built earlier in the facilitation. It is essential that this understanding be translated into anticipatory actions based on the expectation that those in power, and those who are served well by the current system are going to be resistant to radical or revolutionary changes. Consider the chart you created earlier, in step three and try to understand the ways in which power might be used to enhance or sabotage your user-design efforts from within this critical standpoint. As much as it is possible, try to bring to light these power dynamics, discuss them openly with everyone involved, and begin to unearth assumptions that may be paralyzing the organization. Many times the power dynamics are well-recognized by everyone but never spoken of in public. This can lead to extremely negative and disabling organizational behaviors. Opening up these black boxes is dangerous, and potentially explosive, but those explosions can be for good as well as for discomfort—and often even the discomfort is good for the organization in the long run.

This chapter has discussed the powershifts that are necessary for effective user-design and the implications for our personal and professional identities. I have called for facilitators of user-design to engage in the hard work that may be necessary to set aside insecurities and personal blocks which keep us from working most effectively within a user-design context. Implications for instructional design and a series of steps were offered to help guide the user-design facilitator in this new way of approaching instructional design.

> What frightens you most about facilitating UD?
>
> What do you think are your primary insecurities that may need to be addressed?
>
> What do you think are the biggest implications for ID when using UD approaches?
>
> Which facilitator step outlined in this chapter will be most difficult for you? Most difficult for the field?

5

Conflict

In Chapter 4 we talked a good bit about apathy and power relationships. Central to these two constructs is the inherent conflict that is predictable as part of user-design. In many cases, when faced with difficult power dynamics, users would prefer to avoid conflict and opt out through apathy. But in order to really engage the user-design process, the organization will inevitably undergo certain conflicts. One of the most common obstacles that anyone working in user-design is likely to encounter, therefore, is conflict. Conflict can crop up in almost any organization and usually will when engaging in something as different and powerful as user-design. Because user-design asks the powerful to turn their own power on its head, conflict is inevitable. Even if those in power are comfortable enough to engage user-design, those who are being empowered, the users, typically are unfamiliar with active participation in the creation of their own systems of learning and may resist, insisting that this is why they hired an instructional designer. Further, some users may rejoice at the possibilities of user-design, while other users will resent being put in the position of responsibility causing conflict within and among the users themselves. This is to say nothing of the competing political and personal agendas which plague all stakeholder participation schemes.

Thus, conflict is inevitable; it is probably inevitable in all ID projects, but when we maintain control, conflict can be shut down, squelched, or over ruled by the more powerful experts or leaders. This has been the typical way that we have handled conflict in the past, but with user-design, we can no longer approach conflict in the same way at all. It is impossible within the user-design epistemology to simply stamp out conflict, ask those in conflict to settle it on their own, or simply control the situation through power. This way of dealing with power has led to poor change adoption and sustainability in past change and design efforts because users feel controlled and powerless and strike out in passive–aggressive fashion. This is, in my view, at the heart of many of our failures as a field. We have created some excellent designs, figured out fantastic ways to integrate technology into classrooms, and produced elegant online-learning environments. However, many of them meet with failure within the crucible of application and implementation because the users are not adequately empowered to help shape the solutions.

Controlling strategies are ineffectual within a user-design framework and it is essential therefore that the basic values of user-design are carried throughout the entire process. In all that you do and all that occurs it is important to remember that power has shifted, and so some of our old ways of coping with conflict, or processes, or leadership are now in the trash bin. We cannot expect people to act in empowered ways in one moment and then switch that empowerment off in the next moment. Because of the pervasive attitudinal change that user-design requires, conflict in particular must be handled differently than what may feel most intuitively obvious. Understanding this sort of change may best be clarified by considering conflict within real life cases of user-design. In the remainder of this chapter, we will examine the case of user-design as it is employed in a home nursing agency and in the formation of community conversation groups in a public school district. These two cases will present a number of realistic problems inherent in user-design from inefficiency to apathy. We will focus on those issues that created conflict and the ways in which conflict can be productively handled. A more complete discussion of home nursing case can be found in Carr-Chellman, Cuyar, & Breman (1998).

UD IN ORGANIZATIONS: Case Background

This case takes place within a home health care agency and focuses on the design and creation of a more complete understanding among all nurses of the information technology (IT) status in the business and the creation of efficient methods for using IT in home nursing situations. The current economic environment for home health care is marked by managed care and capitated costs. Within the industry, these forces have caused leaders to consider the advantages of IT in the hopes of streamlining costs, reducing staff, and increasing efficiencies in their business.

Home Nursing Agency provides home healthcare across several rural counties, employs 950 people, and delivers a variety of nursing services. Nurses in he home nursing agency complete more than 750,000 home visits each year. Prior to any user-design intervention, the facilitator wanted to find out as much as possible about the current context in the home nursing agency and the way that IT currently fit into the organization. He learned that the home nursing agency used IT primarily to track patient admissions and discharges, provide patient care maps, and perform general secretarial and accounting practices. He also learned that the home nursing agency had an insufficient infrastructure with an understaffed IT department and antiquated hardware. The facilitator learned that on a typical visitation, the nurse would record patient and visit information by hand and then return the information to data entry personnel at the home office. Attitudinally, the nurses were very attached to the paper-based processes and felt that any sort of IT solution to the home visitation process was likely to increase the emotional distance between nurse and patient. One of the things that makes this case so compelling is that it illustrates how flexibly user-design tools can be used within a real-world case of employing user-design on a large-scale IT project and highlights some of the common conflicts that occur in the process.

Organizational User-Design in Process

The home nursing agency user-design process began with leadership agreement and user-design team member selection. The user-design team, with a diversity of roles and representatives, met weekly for more than seven months. Broader stakeholder participation was facilitated through pyramid groups. The project goal was to create a plan to strategically increase the use of IT and consequently reduce home visit costs.

During the weekly meetings, the team engaged in a process following Reigeluth's principles of educational systems design (1993) slightly modified to fit best with the corporate organization. Based on this process, the team assessed the readiness of the context, gained stakeholder participation, selected a whole-business approach for their change effort, and began learning about some IT areas such as data warehousing and point-of-case information systems. Based on these learnings, a common language was constructed and through the extended use of pyramid groups, some fundamental conflicts regarding job security were unearthed. The team identified core business needs and core goals for the team. Based on these needs enabling systems were created including infrastructures, management systems, support systems, and patient care information systems. Once the basic design for the new system was negotiated, the implementation was planned and actualized and then the results were documented and marketed to the rest of the organization.

Discussion of the Organizational User-Design Case

In implementing each of the five strategic information systems plan projects, the home nursing agency utilized the concept of evolutionary prototyping—a hybrid of user-design techniques. Evolutionary prototyping denotes a development approach where information technology systems and the organization are developed simultaneously and integrated (Thoresen, 1984). Floyd (1972) defines evolutionary prototyping as a sequence of cycles: (re)design, (re)implementation, (re)evaluation. The evolutionary prototyping model sketches out the desired outcome with the end user and then constructs a prototype to be used by a selected group of users. Evolutionary prototyping is closely linked to the cooperative design process described in Chapter 3 as it is a hybrid of cooperative prototyping.

Throughout this case there were two primary advantages in employing user-design First, obviously, there was an increased amount of stakeholder participation over traditional design techniques. More traditional approaches such as waterfall or spiral development methodologies limit the decision making to a few individuals. The increased level of participation through the use of user-design and pyramid teams enabled all participants to have a greater voice in the decision making process and helped to create a sense of unity among all levels of the home nursing agency staff throughout the project.

The increased project participation also helped to provide a large degree of project momentum. These inertial tendencies made it difficult for one or two people to monopolize the project direction and related decision-making processes. The project momentum also enabled continued progress despite inherent organizational bureaucracies.

While stakeholder participation and project inertia were advantages, there were also many disadvantages which can help us to understand the conflict inherent in user-design approaches. While the home nursing agency Board of Directors generally considered the project a success, they were also quite cognizant of the elongated period of time required to make and effect decisions. This level of inefficiency was a hindrance to the overall scope of the project. The time required to educate participants about basic terminology and user-design techniques added an additional fifteen days to the project. In the fast moving world of agile organizations, an additional fifteen days is not insignificant. And while it may be that the end result can be justified because of the increases in adoption and buy-in from among users, there is a clear and present threat to the ability of user-design to be effective within organizations in this case. The conflict that is naturally raised as a result of leadership feeling the keen losses of such an inefficiency should not be underestimated. These inefficiencies are perhaps the most obvious disadvantage of employing user-design. The best way to deal with and combat this weakness of inefficiency is to carefully demonstrate both the efficiencies and inefficiencies of employing user-design. This means that a careful explication of return on investment which accounts for both the extra time necessary for user training and negotiation and ultimately design must be balanced with the positive effects of higher adoption rates. It can be difficult to know precisely how much more efficient the adoption process is as a result of user-design and demonstrating that with clear numbers can be even more elusive. However, it is imperative that the UD facilitator identify appropriate variables and measures which can illustrate the ways in which moneys are saved with effective and efficient adoption cycles. There is not a single formula for demonstrating these advantages and disadvantages of investment in dollars and cents, but rather it will be completely dependent on the particular context and up to you, as the facilitator to identify appropriate measures.

In the home nursing agency case, some project participants also saw the loss of traditional power structures within the organization as a disadvantage. The executive management staff had difficulty relinquishing total project authority to the task force. The decisions surrounding project expenses and external consultant selection had to be approved by senior management. This final approval process was seen by some of the project participants as a controlling mechanism for the overall project scope and implementation. As we'll see in the Evergreen case that follows, this is another extremely common drawback from the perspective of the organizational leadership.

In reality, this case illustrates that user participation in design efforts is a very difficult and time consuming task. Entertaining competing ideologies in the process of creating something new can become a protracted process and many teams get bogged down in the details and negotiations, and begin feeling that they are not really doing anything productive or making enough forward progress (Carr, 1994). This scenario constantly resurfaced throughout our project and the facilitation efforts were monopolized with keeping the competing "entities" happy and focused on a common goal of creating a new information system.

The Home Nursing Agency, after engaging in this user-design process, identified a three year strategic information systems strategy. The home nursing agency engaged in action research processes which enabled the participants to learn how to effect change in their own organization on an ongoing basis. While the extent to which this particular case meets the stringent requirements for legitimate user-design initiatives as laid out in Chapter 3, it remains a strong case of many of the hallmarks of user-design. Its primary weakness is that it is more narrowly focused than we would like in any systemic change effort. Just the same, we must recognize that the task force viewed their work in a systemic manner and coordinated with other groups, or spun off additional process action teams to deal with specific systemic implications of their work. However, its charge and the primary focus of its work centered on the creation and implementation of an effective IT solution to home nursing agency's clinical functions. Thus, it is essential that we distinguish here that simply recognizing the systemic nature of the world—that everything affects everything else—is not sufficient to term an effort either systemic or user designed. This discussion of the home nursing agency case gives us an opportunity to highlight the ways in which the case does and does not represent user-design theories.

Pros: Arguing FOR the User-Design Approach.
- Significant stakeholder participation including decision-making powers.
- Shifting power dynamics.
- Active involvement and interaction among users.
- Cooperative prototyping employed.
- Leadership buyin.
- Process was negotiated among all participants.

Cons: Arguing AGAINST the User-Design Approach.
- Some work was accomplished prior to engaging users.
- Facilitators' goals and agendas tended to value buy-in over indigenous users' knowledge.
- Software vendors regarded as experts.

It is clear from this case that stakeholder participation was significant; many in the organization, while not all, were involved in the design process. Users were engaged in the creation of their own prototypes for improved IT usage with home nursing visits and this required certain power shifts which were agreed to by leadership. The meetings indicate an active involvement among users and the process was negotiated throughout all of the meetings. However, the work was not completely bounded by the user-design process. And this meant that there was most certainly somewhat of a feeling that the users were not critical to the entire process which argues against considering this as true user-design. Knowing with any certainty what the cultural impact of this prework was on the entire process is impossible. It may be it had little or no impact, or it may have undermined the user-design process from its inception. Because the software vendor was seen as an expert and was relied on by users as an expert, the UD process was compromised in ways that tended to undermine the sense of respect for indigenous knowledge. And the facilitator tended to value buy-in more than indigenous knowledge.

Implications of the Home Nursing Agency Case

For-profit companies wishing to undertake user-design should carefully consider the role of *leadership* when initiating user-design efforts. Without the buy-in and permission of leadership, the process will stall very early on as the users will not sacrifice their valuable time in engagement with deep design issues without first being convinced that their voices will be heeded in decision-making. A certain amount of justified user cynicism must be overcome, after all. Because of this fact, traditional hierarchically based leadership structures need to be abandoned in order to truly empower users in the design process. Alternative leadership models, and particularly those models that tend toward transformational leadership styles (Burns, 1978) are preferred for user-design efforts. A more specific discussion of leadership for user-design will follow in the next chapter. But, the home nursing agency case obviously brings the role of leadership into sharp focus.

Another related issue which surfaces as a result of this case is the role of *expertise*. The software vendors were seen by users as experts in this case. Recall that users will seek out expertise and leaders rather than taking on active and responsible design and decision-making roles themselves because this experience is such unfamiliar territory. While the design implications of the theory have already been set out (Carr, 1997), user-design facilitators are faced with the difficult question, "How can you move a group toward new design ideas without an overreliance on expertise or leadership?" Rogers (1995) has argued that change and diffusion are aided by facilitation among most similar groups—called *homophily*. However, he also points out that something new and therefore different or heterophilic needs to be present for an innovation to exist. The difficulty is how to equalize these two forces which exist naturally in all groups. In user-design the balance between the energy of expertise in bringing new ideas to the design team and the debilitating effects of expertise as a crutch which can co-opt user's innate design abilities.

Finally, in terms of design, this case suggests that we need to improve processes and tools which help the uninitiated individual user to effectively engage in the process of designing their own systems. The process of design can be presented as a difficult, jargon laden, technical enterprise, or as a simple problem solving activity. To the degree that we, as facilitators and designers, can express the activity in clear language which does not somehow jeopardize the validity of the design practice, or oversimplify it, we will be able to continuously engage new users in design. But, in addition, we need tools such as cooperative prototyping and other innovative approaches which are slowly emerging from HCI literatures that can be adapted for the purposes of UD.

UD IN SCHOOLS: Case Background

This second case takes place in a large suburban public school district on the outskirts of a large metropolitan center in the northeast. This case focuses on the creation of user-design conversation teams primarily to consider several curricular innovations and build community relations in a school district that had some historical difficulties with their ongoing stakeholder participation. This case is particularly useful in terms of talking about conflict because there was significant conflict engendered among and between leadership and stakeholders.

Evergreen is a school district with approximately 2500 students and 200 faculty with an additional 75 staff. The district has been very traditional in its change processes however, it has embraced many new innovations through the years and has had many waves of change throughout the past two decades as has been the case with many similar districts. Evergreen is located in a relatively wealthy suburb with large houses and excellent shopping districts. In most cases, teachers do not earn enough on their own to be able to live in housing that is readily available within the district. While some, particularly at the elementary level, are second income earners and are able to live in the district, most teachers and staff live in surrounding areas with more plentiful affordable housing.

The facilitator in this case was invited in by one of her students who was intrigued by participative action research, qualitative research methods, and educational systems design as studied in her classes at the local university. The facilitator learned that the school district suffered from what some term the "Christmas Tree" metaphor of change in that many ornaments had been hung on the tree through various change efforts, but system wide, organized, coordinated change had not been forthcoming.

School-Based User-Design in Process

The Evergreen user-design process began with a somewhat protracted negotiation with leadership around issues of control and decision making shortly followed by user-design team member selection. Whereas in the home nursing agency case this was a relatively straightforward part of the process, it was much more difficult and time consuming in the Evergreen case. As is the case in many school-based exemplars of traditional stakeholder participation, first identifying the stakeholders and gaining serious buy-in from them is very difficult. In a context where all participants are volunteers, ensuring any sense of ongoing commitment is a significant obstacle. And this obstacle can derail a UD project before it even gets off the ground. In many cases, stakeholders are reluctant to sacrifice their precious few hours of volunteer time unless they are sure there will be a clear and significant impact on their own child, or on the short term future of their school district at most. Lofty goals of improving education for future generations are difficult to sell to an all-volunteer user-design team.

In Evergreen's case, the team consisted of 16 parents, high school students, and community members who met each month for eight months to discuss a variety of issues going on in the school and design potential solutions. There were no pyramid groups in this case, but it was emphasized to the team that it would be helpful for the process if they could gather additional input even if they disagreed with it, and share it (not necessarily represent it) to the team during meetings. The project goal was primarily to provide insights and suggestions on curricular designs for the Evergreen School District. The leadership saw this as a two-way conversation in which they could share what they had learned or knew about "best practices" while they could learn from stakeholders what their perspectives on the ongoing curricular improvements were. The charge was not limited to a single school or a single curricular innovation. It is important, however, to distinguish this from a team that was merely put together to provide input on the job satisfaction of the district from stakeholders. The team was created as a user-design team, expected to engage in design activities, and to create new solutions to curricular problems presented by the team as well as the leadership.

During the monthly meetings, the team engaged in very open design conversations with very little formal structure. Unlike the home nursing agency case, where Reigeluth's model was followed as faithfully as possible, the Evergreen user-design team did not subscribe to any specific user-design model, but rather employed pieces of several design approaches including Reigeluth's model (1993), Banathy's model (1992) and some work by Carr-Chellman (1998). Thus, the team selected a whole-district approach for their change efforts and began learning about some curricular innovations that were taking place across several schools. At the same moment they learned about user-design and design processes and models and employed those parts of design approaches that seemed to fit best and feel most appropriate to them as a team.

Discussion of the School-Based User-Design Case

Evergreen experienced a fairly typical profile of recent attempts at UD, or more generally, stakeholder participation. Members of the discussion team came and went over the eight months duration of the project. This, naturally, creates a certain amount of aggravation among both leaders and faithful participants. When members join a meeting and participate and then do not return for two or three months, it is very difficult to re-engage. In addition, it is common that when a participant returns after such a long absence, a good deal of time is spent by the entire team briefing the group on progress made in their absence. Brief reviews are possible and appropriate for all teams after a month has passed between discussions. However, when conversations are essentially re-accomplished and topics are discussed at some length after having previously been settled, it is frustrating to other team members and leaders. The team can flounder and feel they are not making any progress if this is chronically repeated meeting after meeting. And in Evergreen's case, the fifth and seventh meetings were particularly illustrative of this source of conflict among user-design team members.

The second very common problem for Evergreen was the dominance in membership by particular populations. It is quite common for membership in teams of this sort to be primarily or even solely majority population mothers. It is all too common for teams of this sort to suffer from little or no representation of minority populations, fathers, and other community members such as business owners, pastors, social service agents, senior citizens, taxpayer groups, etc. Naturally, these limitations on membership have unseen restrictive impacts on the user-design team in terms of the ideas generated as well as the levels of possible acceptance of new designs. Gaining and sustaining broad membership has been a traditional stakeholder participation obstacle for decades. In user-design it becomes even more essential because the activity is not simply gaining input, but creating new systems. The time consuming nature of user-design is balanced by increased efficiencies in adoption and implementation, but these efficiencies can easily be lost should the team be made up of too narrow a slice of the community. For Evergreen, the membership was typical of most other school community stakeholder teams and so they specifically sought out new parents from minority opinion positions as well as minority populations more traditionally defined (by race or income). The leaders at Evergreen also invited specific representatives from the community and had some success at gaining ongoing support and membership from several community members who would probably not otherwise have been involved. Sometimes, just being asked is sufficient to start the user-design process down a more productive road. However, because of the spotty attendance of all participants, the team became frustrated after the first year of meetings.

The third and most important area of conflict that emerged during the course of this case was a disagreement within the leadership. The superintendent was involved in the user-design conversation groups from the start of the project—but from a distance. That is, he was always informed about the project and updated from time to time about the progress being made. However, the superintendent never attended the meetings or took an active role in the team's formation or operations. This definitely communicated a certain message to the participants. While it was not clear whether the user-design team felt supported in a limited way, or unimportant to the central functioning of the school, this benign neglect ultimately did have an impact on the perception of team members in terms of the importance of the team itself. You may feel as if leaders can't win in user-design. Either they are too hands-on and can be seen as manipulative, or too hands-off and therefore unsupportive. Finding a proper balance between disinterest and control really calls for active participation of leaders who invest their time, but do not expect to direct the effort in their own direction.

On the other hand, the assistant superintendent and the director of curriculum were actively supportive of the group—attending meetings and sharing information as well as listening to the conversations that emerged. The active school administration, however, struggled mightily with the idea that users would be empowered in the system. So, they participated but were reluctant to relinquish power. This is the most common shadow self of the benign neglect shown by top leadership. One administrator said:

> It's our job to educate these people (meaning the conversation group members) about what we know as educators. We know what's best from best practices research, we have to find ways to let them know about this so then they'll be more likely to accept the new math program. We know it's what's best for kids.

This sort of reluctance to really take seriously the concerns of users, parents, learners, and community members as they opposed a particular curriculum innovation highlights one of the most difficult obstacles and conflicts anyone embarking on user-design activities is likely to face. It speaks to the primary conflict; the disagreements over how power will be distributed.

This case clearly illustrates the difficulties encountered in UD in schools which are community/social agents of change. Schools are particularly difficult institutions in that they do not have a clear charge from the community or society—we have not yet agreed as a society if schools are to be community centers, education centers, even babysitters. user-design does not complicate this problem; it brings it to the surface in ways that are healthy and exciting, but also in ways that are time consuming and conflict-ridden.

Evergreen school district, like many school districts who engage the stakeholder participation or user-design approach, allowed the initiatives to fade over a period of time. When administrators realized that this was about power shifts and what that really meant for their own practices, and when participants realized that they were not going to be accorded any real power, the conversation groups floundered.

In similar fashion to our consideration of the home nursing agency case, we can see ways in which this initiative is or is not user-design from the strictest definition of the term.

Pros: Arguing FOR the User-Design approach.
- Broad stakeholder participation.
- Active involvement and interaction among users.
- Initial leadership buy-in.
- Process was negotiated among all participants.

Cons: Arguing AGAINST the User-Design approach.
- Power shifts were not accorded to users.
- Indigenous knowledge was not central to case.
- Administration maintained expertise.

It was definitely the case that Evergreen wanted to find a broad stakeholder base for their community conversation groups. Even the term they used invited everyone to the table. They were expecting and hoping for active involvement and inspired a great deal of interaction among the users through ongoing dialogue. There was some initial steps in the right direction in terms of meeting the lofty goals of true user-design such as a socially negotiated process and initial leadership buy-in. However, the process really fell short of the goals when the power shifts were resisted mightily by the leaders. Indigenous knowledge, then, was not respected and the administration held onto power and expertise as a model for continued school change.

Implications of the Evergreen Case

Schools of all sorts, public, private, charters considering a user-design approach will most clearly have to consider quite carefully the extent to which power issues are really open possibilities for them. Unlike for-profit companies, the school tends to have multiple goals and a lack of agreement on their primary business or purpose. They also are social institutions and because of this are more accountable to community needs and desires. So, for many reasons, user-design seems a logical choice for schools. However, the conflicts that arise when power is shifted in any community are not to be underestimated, and schools need to carefully consider how much power they are able to share or shift. If a user-design effort is taken on and subsequently aborted, participants may be disappointed or bitter toward the school and will be hard-pressed to participate in future similar efforts in almost any public sphere.

However, the case for the implementation of user-design in school change communities is very compelling. While our government seeks out generalizable solutions that can be diffused throughout the entire public education system in the U.S., there are those among us who believe that local solutions are the only really viable possibilities for systemic change in schools. While policy advocates do invoke the term *systemic change*, mandated change cannot truly engage users in powerful ways because they force a particular solution on the users in advance. The job, then, of the change agent is merely to find the best ways to get users to do what those who are mandating the change want them to do evoking Rogers-like perspectives on the change process. The problem is that schools are each too different to be able to use others' solutions. We may have processes, theories, or approaches that are worth using with a variety of school communities in an effort to assist with the process, but the products need to come from the local setting itself. And generally, when processes are adopted wholesale to a new community, they suffer from the NIH (not invented here) syndrome among users. More transition to user involvement in the process but some guidance for the process is possible; that is, after all, what this book is about. This understanding is key to user-design approaches. The approach is scalable, but the solutions are not. As MacDonald (1971) points out:

> No two schools are sufficiently alike in their circumstances that prescriptions of curricular action can adequately supplant the judgment of people in them. Historical/evolutionary differences alone make the innovation gap a variable that has significance for decision making. (p. 183)

The other primary conflict that is likely to arise in any school-based user-design, or for that matter any volunteer user-design effort, is the level of *commitment* from among the users and the agreements they have in terms of time and effort. User-design is not a passive activity. Indeed, it is essential that all users really engage the process of creation and design and this requires time, commitment, and active participation in ways that are very unfamiliar to most who have a history in the PTA/PTO, or other traditional school committees. Gaining this high level of commitment and agreement from participants in user-design teams is essential if the team is to function properly, because the problems with variable attendance should not be underestimated. This was another of the conflicts that arose in the Evergreen case that illustrates the sorts of issues school administrators and school communities must address prior to undertaking a user-design approach in their next school innovation.

As with the home nursing agency case, the processes and tools utilized in user-design for school change need to be dramatically improved. There are some promising innovations such as open space technologies (Owen, 1997) and preferred futuring (Lippett, 1998) which help large groups of people create new ideas and design their own future. However, use of these and similar technologies in the design of the new World Trade Center architecture highlights many of the primary conflicts that these tools engender (Filler, 2005).

Discovering truly powerful tools for use in school communities will be essential to the future successes of user-design in schools as well as organizations and community building of all sorts. And careful attention to these issues of power, commitment, and tool development is essential to the development of UD as a method for school change.

ISSUES OF CONFLICT FROM THESE CASES

Let us now focus more specifically on conflict. Conflict occurs whenever we have diverse agendas playing themselves out in the same system. We can spot conflict all over the world. A current example is the U.S. social security system. The problem is actually pretty clear—the social security taxes paid by generations past have been spent by lawmakers for a variety of initiatives to support many social programs of merit. The understanding has always been that current workers will support the retirees of a particular generation, but now we face a crisis of demographics. We have known for some decades that the number of workers may not support the number of retirees, but with improved healthcare and increases in longevity, retirees are living longer and the system is in crisis. There are actually a number of relatively straightforward solutions. For example, we can delay retirement, we can increase the amount that current workers pay in, we can take monies from other programs. So why haven't we fixed the problem, and why do we create elaborate and complicated possible solutions? Basically, it is because we, as a society, are in conflict over the crisis. Workers under a certain age are not interested in paying more taxes; retirees and workers over a certain age have a significant stake in keeping the system as it is to ensure their benefits, which they've been guaranteed by the government; support them in their old age, recipients of other social program benefits are loathe to consider their benefits being put on the chopping block to benefit the social security program. Perhaps in order to disguise the conflicts inherent in any solution, lawmakers and leaders suggest complicated ideas and plans for solving the crisis. If you look at the systems in your own life you'll see the conflicts that are just beneath the surface. Conflict is simply a part of life. We each see the world differently and we have different wants and needs; this is what conflict is borne of. And thank goodness we *DO* have different perspectives how boring and unidimensional our world would be if we, in fact, all agreed. And, how limited would be our solutions to problems if we never had conflicting agendas! It is in this crucible of conflict that great ideas are created and specific solutions designed.

HOW TO NAVIGATE THE CONFLICT

Thus, the problem is not conflict itself; in fact, conflict is a primary source of energy in design. When we try to eliminate conflict between people, we are actually robbing the process of some of its elemental energies necessary for creation. An examination of the concepts of conflict resolution and peacekeeping indicates a tendency toward consensus building. However, consensus building has been strongly criticized by many who point out that consensus tends to build solutions that do not substantively honor anyone's agendas and sometimes finding compromises can weaken the final approach.

Thus, typical approaches to conflict such as consensus and compromise have limitations. Certainly in some cases, consensus and compromise can be powerful tools, particularly when the parties are relatively close together in their basic underlying values. However, in many cases, when parties disagree stridently, we don't need to eliminate conflict we need to handle it and harness it. So, how do we do this without alienating whole groups of users who disagree with one another? I've run across a similar set of problems when writing about religious diversity in the workplace with my husband (Carr-Chellman & Carr-Chellman, 2003).

What we proposed was that awareness of others' religious differences is a first step; understanding and empathizing with others' perspectives is a good place to begin but a terrible place to end. Rather it is essential that this awareness build into significant dialogue amongst all stakeholders. In a similar vein, recognizing the centrality of values in this endeavor, Ray Horn and I (Horn & Carr, 2000) wrote, "Moral purpose can provide the motivation to work with others in developing a shared vision and implementing change that enhances the vision of the community" (p. 6).

The creation of this shared vision can only happen through conversation and communication with the community. Horn & Carr (2000) continue, "Moral purpose differs from context to context and must be determined through dialogue that includes all members of the community" (p. 6). In the world of UD we call this "design conversation," which goes beyond what we typically think of as conversation. Most of the time when we are in conflict; particularly when the conflicts are fundamental and speak to the very underlying values that are being recreated, there is a tendency to set up discussions.

We see town meetings and open forums available for public comment. But, these are rarely any kind of real communication. The participants enter the arena with three-minute speeches pre-prepared, rehearsed, and complete. There is no room for changing minds or for real engaged communication. Even when we meet in small user-design teams, we too often see discussion in which one person hits or percusses—others with their ideas, trying to convince them that their view is right and others' are wrong. The International Society for the Systems Sciences (ISSS) has recognized, in their Primer Project, the nature of dialogue as both important and too often mere shadows of serious engagement:

> Dialogue is essential for understanding cultures and subcultures in the emerging global village. Boundary-spanning dialogue across disciplines and civilizations, if conducted wisely, can generate democratic agreement on the courses we must pursue to create *agoras* and avoid Big Brother. Thus, the ability to engage in dialogue becomes one of the most fundamental and most needed human capabilities. Dialogue becomes a central component of any model of conscious evolution. (ISSS, 2003, p. 1–2)

> In today's complex world, remote forms of constitutional representation pass for democracy. Traditional democratic forums, being cumbersome and unfocused, can no longer guide effective civic action. Town-hall meetings and conferences (whether international or academic) are pale vestiges of agoras. They are staged events that deal superficially and/or narrowly with complex processes. (ISSS, 2003, p. 3)

CONFLICT 73

In handling conflict within user-design, it is imperative that we employ the best tactics of design conversation and that we prepare all participants for powerful, yet potentially limited, experiences. What I mean by this is that those who are engaged in user-design will have high hopes that their words will be heard, and this means they expect, often unrealistically, to get their way. In any team, there will be conflicts, and when things as fundamental as public schools or communities, or systems of human learning are being designed, the conflicts will be substantial. This means that not everyone will get their way, because that is simply impossible.

In effective design conversations, there are two important things that must be established. First, the most powerful doesn't always win. This is critical. If those in power always win, then those who have no power will disengage, and they are right to do so. The system is not respecting them or their knowledge, so why waste valuable time? Second, all participants must understand that since their views are not all going to be the same, diversity may occur and may be accommodated rather than stamped out. This is not always an easy thing. When we think we are right, it becomes like scaling a mountain to admit someone else may also be right if their view is diametrically opposed to our own. No, in this age of *Crossfire* and call-in talk radio, we have become almost immune to the raised voices and vitriolic rantings of people on different sides of issues. We have come to accept it as normal communication, but it is not; and, it is not generative communication in any way. It would be unheard of for us to hear Rush Limbaugh or Bill O'Reilly accommodating viewpoints that oppose their own into some plans for real action. Yet in a pluralistic society, we essentially agree to allow or accommodate some level of difference among our positions on key issues.

Design conversation calls us to disrupt our usual communication patterns and those that are modeled for us on a daily basis in order to create effective user-design efforts. How can we initiate and sustain design conversation as a method for dealing with conflicts inherent in the user-design process? First, we must understand what design conversation is. For a more extended explanation, see Jenlink & Carr (1996) in which a typology of conversations from dialectic and discussion on one end through dialogue and design conversation on the other end are discussed. Once we have a clear understanding of the differences among conversation types, we can more easily recognize when we are "in" design conversation, which feels quite different from other forms of conversation. When you find yourself engaged in conversation that centers on creating something new, you will feel a different sort of energy flow than when you are engaged in discussion or debate. This sort of conversation accommodates variety much more easily than more oppositional types of communication.

Many of us have been in design conversations in our everyday lives. I recall the creation of a kitchen space that my husband and I worked through for hours and hours in the renovation of our home. The space was odd, with many doors and architectural constraints. I remember him saying, "It was like real labor." But, it was also extremely energizing because we were engaged in something real and tangible with results that were going to become reality after all of our talk and sketching. And while we agreed fundamentally on the basic values we wanted to express with the space (functional yet elegant), we also were able to suspend our own visions long enough to listen and really hear the ideas of the other. We also had complete confidence that we each had power—it was not as the contractors joked that "the boss" was always the wife. In the end, the kitchen emerged as a joint project; neither of us can really remember what our original ideas looked like or who came up with what innovations that we adopted. It was a real crucible of conversation—a design conversation.

But, it will not be so simple to harness this energy in the creation of human learning systems with a large number of users. Fortunately, there are many methods to inspire design conversations, some of which are outlined fully in Jenlink et al.'s (1998) work on facilitating systemic change. In that work, we suggest a number of specific strategies to help users break out of their limited mindsets without engendering a more patronizing tone of "educating" the users—implying that their perspectives are not as informed as those who are doing the educating. Experiences such as field trips to see very different business approaches or school designs, videotapes, lectures, readings (where literacy is firmly established), and even graphical images are all good seeds for germinating true design conversation among user-design teams.

Once the design conversation process is established and inspired once the participants are prepared for diversity, accommodation, and power shifts, the users can fully engage in the creation of their own systems of human learning Often we can find a basic agreement in terms of broad goals or values; it is in the specifics that we begin to come into more strident conflict. As the old saying goes, "the devil is in the details." There are many ways of trying to deal with diversity of opinion on how to implement specific changes. For example, school-within-a-school systems are possible whereby a variety of instantiations of a similar set of values can be lived out by different groups within a single, large physical school plant. Charter schools have also allowed districts to embrace a variety of ways of approaching learning that fall within the broad values they establish. This is a good example of accommodation.

While I personally may not be in favor of for-profit corporations taking over schools in order to show efficient test score advances, if allowing for that within the public schools means which I can create a school that best meets the needs of my children and others like them, perhaps it is a trade that is acceptable. This does not lead to a sense of moral relativism, however. Rather, as Buber (1992) suggests, such diversity is the foundation upon which constant change releases energy into the system. Without change, we stagnate and even die:

> The domination of any one element over the others, and the potential annihilation of any such element by others, may produce just routine organizational or structural change, or it may lead to the stagnation or even to the demise of a society or culture. (p. 11)

Thus, within all of this debate which we use to convince others' so that we can all agree—within *that* lies our extinction. This is not moral relativism; rather it is the celebration of values and the accommodation of diversity in ways that transcend political correctness and truly recognize the value that this diversity brings to our community.

CHAPTER SUMMARY AND THOUGHT QUESTIONS

Handling conflict can be extremely difficult. And, if it is done badly, a facilitator may be quickly escorted to the nearest exit—perhaps with sighs of relief on both sides of the door. However, when conflict is handled properly, with a healthy attitude and embraced as an energy source rather than a drain or a fearsome boogeyman to be avoided at all costs, it can be the impetus for the most exciting of designs. In this chapter I have covered a number of necessary steps for handling conflict in ways that lead to generative design conversations:

- Build awareness among users of diverse agendas.
- Engage users in dialogue and design conversation (as opposed to debate or discussion).
- Build an expectation that the most powerful do not always win.
- Allow for accommodation of diverse ideas.
- Use compromise and consensus only as appropriate but avoid overreliance.

- Embrace conflict; don't avoid it.
- Understand different conversation types.
- Distinguish when users are in design conversation and encourage it.
- Inspire design conversation with stimulus materials (field trips, videos, etc.).

These steps certainly won't guarantee that conflict can be harnessed or that an enormous blow up won't occur and be publicly aired on the pages of the local newspaper, radio airwaves, or community television. But they are a good start in our quest for engaging in a user-design that really works.

> What do you believe will be your own biggest personal challenge in handling conflict in user-design?
>
> How do you distinguish between design conversation and other types of communication?
>
> Can you give an example of design conversation from your own experience?
>
> In a context where you would like to apply user-design, can you anticipate some sources of conflict and ways to handle that conflict?
>
> Which of the steps listed above will be most difficult for you? For your organization?

6

Leadership

Perhaps this section of the book should be called, "The care and feeding of leaders" because most user-design facilitators are unlikely to be leaders themselves. Sometimes the initiation for a new design project comes from leadership and sometimes from the grassroots. In either case, however, unless you are in a context of extremely strong union constituents, such as is commonly the case in Scandinavian countries, you are likely to be facilitating a change process in which the leader has never heard of user-design.

Because of unfamiliarity and also because of a fear of shared power, most leaders will initially shy away from engaging in user-design efforts. "The traditional model of change hasn't worked," they may admit, "but this idea is too radical and will turn all my power over to the employees (or learners or users)." Thus, many times, the first step in working with leadership for the user-design facilitator is to convince them of the merits of such an approach. Recall that in Chapter 1 it was made clear that user-design is not appropriate in all cases, and that care should be taken to ensure alignment between the goals of the project and user-design as a match. If that match is missing, a serious disservice is done to the leadership at the outset and trust will be compromised.

Once it has been established by you, as the user-design facilitator, that this is an appropriate approach, the next step is to clearly illustrate what the process is about and how the process can benefit the organization. These two things should be clearly communicated in written documents to the leaders of the organization. A one or two page administrative summary should lay out the basics and a lengthier document can detail the cost–benefit analyses as well as specific plans for how user-design might be carried out in the given context.

Now understanding leaders themselves is an important first step to learning how best to work with leaders on a user-design approach. First, it is my view that good leaders are actually pretty transparent. Good leaders are not really the charismatic out-in-front people that are often lauded in the press. Really good leaders recognize talented workers and get the heck out of their way. They encourage autonomy and reward initiative. They are not threatened by talent; rather, they do all they can to foster better and better talent. I was recently in just such a leader's office. He had a little Xeroxed sign above his desk that read:

> Blowing out another's candle
> Will not make yours shine any brighter

I was very impressed by the sentiment and asked him about it. He told me that usually the people who sit where I was seated really needed to read that sign and recognize that their insecurities were simply causing darkness. Good leaders already instinctively know this. They give credit wherever it is due, and as a result they inspire greatness in their followers. All leaders display behaviors which impact their followers (Kelley, 1992); good leaders display behaviors which create healthy organizations filled with workers who love what they do. Insecure leaders, on the other hand, are much more likely to follow behavioral patterns that may get their faces on the evening news as outstanding leaders, but inside their organizations life is pretty different. If you doubt this, consider the case of Microsoft. Certainly Bill Gates has been an inspiring leader, and his face has been plastered on every money magazine known to mankind as the richest man around. But look deeper into the culture of Microsoft and his own history and you'll see a different story as told in *Microserfs* (Coupland, 1995) or the film *The Pirates of Silicon Valley* (Burke, 1999). So, while, the business schools study the successes of leaders like Bill Gates, the really excellent leaders are the ones you never hear about because they are merely transparent conduits through which lots of good work gets done.

A truly good leader is like a *Koi*. When you are considering an organization for user-design, be on the lookout for Koi. A Koi is a fish that comes originally from China and was brought to Japan where it eventually made its way to the West. Koi are colorful fish that are relaxing to watch and very resilient. This is why they make great additions to backyard ponds. They are non-aggressive toward other fish and are mostly motivated by food. They have no teeth, but they have boney plates deep in their throats. The Koi is a symbol of strength in Confucianism. Good leaders are like Koi fish; they are relaxing to watch—they do the job without a lot of flash or fuss. They don't complain, and those who work for them relax around them. They are resilient—they do not let the little things get them down. Koi do not eat other fish, nor are they aggressive, a truly good leader gets out of the way and enables others, they do not take good ideas or productivity of others on as their own success or failure. They are motivated by food—they keep incentives simple, and they know what counts in their organization. They may not have teeth, but those boney plates deep inside are very powerful just the same.

If you are in an organization with a Koi leader, you'll have ample opportunities to engage in user-design. But there are many different types of leaders, and Koi are rather rare fish after all. Types of leaders have been discussed many places (e.g., Kelley, 1992 Burns 1978), and general categories or types of leaders range from autocratic at one end of the spectrum to transformational at the other. In general, studies have found that the more autocratic the leadership, the less successful have stakeholder approaches been. However, there are ways to try to help autocratic leaders to move toward more open leadership styles that allow for the power of user-design in their organization. And, regardless of the type of leader who heads your organization, there will be conflict, so be sure that you are intimately familiar with the guidelines in Chapter 5 on handling conflict. However, there are some special measures that should be taken when dealing with leaders in a user-design context.

First, but probably not primary, the cost–benefits need to be clearly explained to all leaders involved in negotiating the implementation of a user-design process. If leadership is caught unaware of the high costs associated with the user-design approach, particularly in the early phases, then the frustration that may be voiced by some of the users during the process along with the costs may overwhelm the viability of the process and shut the whole thing down prematurely. Instead, make absolutely sure that a clear delineation of costs associated with the process during different stages is communicated. Make sure the leadership understands that these costs are considerably outweighed by the benefits which come from smoother adoption and implementation of innovations and if at all possible, illustrate these benefits and costs for them with dollar signs. For many leaders the adage "show me the money" will work in this instance. As has been the hallmark of instructional design for decades, cost–benefits analyses are compelling counters to the higher up front costs of training design and development. Following in this tradition, user-design facilitators should take it as their obligation to communicate the costs and benefits before embarking on user-design.

Another potentially less mathematically obvious benefit is the organizational harmony that will result from user-design longitudinally.

While in the short term it will be the case that user-design creates certain disharmonies or conflicts within the organization, over the long term, user-design's conflict is seen as healthy and normal—as a source of energy and innovation. As users are able to look back over a successful user-design project, they can feel pride in their accomplishments and power over their own destinies. Mandated change robs users, followers, learners, and stakeholders of their power and ignores the wealth of indigenous knowledge that they have to share. While it may be threatening to some, most leaders can understand and appreciate this fact. It is the job of the user-design facilitator to ensure that the leaders have at least considered this additional benefit that may not be quantifiable.

Of course the resulting systems of human learning that are created as a result of user-design are much more likely to produce their desired outcomes. User-design simply creates the right milieu for great ideas and therefore better products, innovations, and designs. Naturally, this will be an important part of any cost–benefit analysis that will be very difficult to quantify because it is impossible to know, in advance, with any certainty what ideas are likely to emerge from the users. Therefore, predicting the pay off of a great idea that hasn't yet been discovered is going to be very difficult to communicate to leadership, but the user-design facilitator should at least try to share their thoughts on this with leaders in advance.

Once the leaders have bought in and agreed to a user-design approach, how should you as a user-design facilitator involve the leaders themselves in the process? There are two primary strategies, and perhaps some variations on these approaches. First, you can follow the fairly traditional plan of keeping leadership informed with written updates. If your team agrees to create meeting minutes regardless of leadership needs, then this might not create much extra work; if not, then there is a fair amount of additional work added to your plate by agreeing to timely written updates for leadership. But of more importance here, keeping the leadership out of the nitty gritty user-design meetings, may create a certain sense in which the leadership is above users in their own involvement. This approach, then, can create a sense in which users feel that leaders have veto power over their ideas. In addition, keeping the leadership out of the meetings themselves misses an opportunity for the users to hear more from leadership about their perspective or constraints that they may not have considered. This can be a double-edged sword because when leaders are included in meetings, they may be accorded certain powers naturally, which can create negative or less than powerful user voices.

The other alternative, naturally, is to have the leadership involved directly by being full members of the user-design team. As suggested above, this offers an opportunity for sharing leadership perspectives which will be critical to the eventual adoption of user-design ideas. It also gives the leadership a clear idea of the user voice and the level of work entailed in user-design processes. It is naïve, however, to assume that the leadership can simply be involved in the user-design meetings without having a significant impact on the power dynamics already inherent in the organization. Thus, it is imperative that you carefully prepare the team members and leadership for the kinds of conflict and opportunities that will arise as a result of their direct participation.

Given all of these pros and cons for each approach, I generally feel that it is better to include the leaders directly in the meetings despite some of the potential drawbacks. There are probably alternatives that lay between these two options, such as participation by leadership as ex officio members. But basically, these alternatives cannot entirely mitigate the pros and cons of these two extremes and so I leave it to you to determine which level of involvement makes the most sense for the organization that you are working with. This decision should be based on the openness of the organization to user-design processes, the leadership style, and willingness to participate, and levels of trust already present among the users.

Another issue when involving leadership is how many leaders to involve and which leaders to involve. Certainly you do not want to have more leaders than users, unless that reflects your organization. Having more leaders indicates that they have a stronger voice or a final vote. Here at Penn State we have a faculty governing council—the Faculty Senate. As a non-union university, the Senate is the closest thing we have to faculty power, but the body has almost 10–15% presidential appointments who are typically administrators and several additional faculty senators who are currently occupying administrative posts, though the Senate considers them faculty. This certainly changes the dynamics of the whole organization. Close votes have been swung by administration alone. So, you have to be cautious about the sheer number of leaders who are involved when compared to the number of users. In addition, you want to be sure that you involve the leaders who are most stridently for and against user-design. That may sound counter-intuitive; why would you want to involve those who disagree with the approach? In any organization, voices of dissent are never completely silent. As is the case with conflict, it is better to deal with these problems up front rather than allowing feelings to fester and rumors to spread throughout the organization. So, in general, it's good to have a variety of perspectives across leadership represented within the user-design team.

Consulting with any team members, particularly leaders of the organization or leaders within the user-design team, should be approached with a certain amount of caution as well. Between meetings there will be work that has to be done, and it will be important to encourage users to persevere in various design tasks. This will require a certain amount of facilitator communication or consultation between meetings. However, this can create perceptions of favoritism or unequal power across the user-design team or the entire organization. Thus, it is important that you consider the rumor mill in all such relationship building. Do not avoid building strong relationships with all users and all members of the design team, this is essential for success, but realize that every move you make will reflect on the process, and be aware of how your actions might be read or misread by those within the organization—particularly those who may not be supportive of the overall approach.

Finally, as you deal with the user-design process, whether it be in a school, business, or nonprofit organization, there will be opportunities to mark significant advances or milestones. It is important to celebrate these moments and highlight the specific outcomes and practical implications of the user-design teamwork. The process of user-design is particularly lengthy and may be frustrating to the user-design team as well as the entire organization because it does not always proceed quickly or smoothly. This frustration can be partially ameliorated by reports of any advances and concrete changes that result from the team's work. This can be particularly important to results-oriented leaders who may be watching that bottom-line. As a facilitator, you will want to issue such progress reports often, but avoid making too many such reports that may be seen as a waste of time or resources. Base your decision on when to offer updates to the organization, leaders, and team members on organizational standards. If it is typical to update once a month, you can do that, although if you see in advance that an update is not likely to report significant advances, it may be better to space out the updates more infrequently but have more important information to share each time. Make sure to negotiate a fairly open schedule for reports among the team and leadership up front. The user-design process is uniquely time intensive and so it will be important that everyone understands this at the outset. Negotiating a slower or more open-ended schedule for reporting progress can help to impress this difference on all stakeholders early in the process. It can also be very useful to consider celebrating the transition from design to implementation with a clear recognition that shifts are happening within the organization and that this will result in a different kind of activity—one of sustaining, maintaining rather than creating.

LEADERSHIP 83

CHAPTER SUMMARY AND THOUGHT QUESTIONS

The care and feeding of leaders, as I opened the chapter considering, is not an easy task. And it can be daunting to think that in the midst of the hard and important work of creating new systems of human learning with the users themselves, you also have to carefully consider organizational culture and how to best work with leaders to generate positive energy throughout the organization and for the duration of the project itself. There are many suggestions for user-design facilitators:

- Maintain clear communication with leadership
- Make sure leaders are clear about the what user-design is and how it can benefit the organization
- Understand the type of leader your organization has and the potential impact on the user-design process
- Illustrate the costs and benefits clearly and honestly up front before beginning any user-design initiative
- Highlight less obvious benefits such as organizational harmony, user pride, better product and process ideas
- Involve leaders directly on the user-design team if at all possible
- Consult with leaders and other team members between meetings cautiously
- Negotiate open-ended progress reporting

Engaging leaders in the process of user-design is perhaps one of the most difficult negotiation points within the user-design approach. Gaining their participation and positive endorsement is essential for any user-design effort. Without leadership consent, the process becomes more adversarial than is realistic or desired. Ultimately, the user-design approach hopes to open the power dynamics within the organization to allow learners to create their own systems of human learning. Bodker, Gronbaek and Kyng (1993) write of this:

> In most organizations, some groups have more power and resources than others. Those who have the most power and resources are usually management, not the end-users. To help users get a forum where they can take an active part in design means to set up situations where they can act according to their own interests and rules, and not simply according to those of their managers. (p. 172)

The idea here is not to encourage users to overthrow the leadership, but rather to avoid such revolutionary change by encouraging true, honest empowerment of users to create what they feel will most help them in their own work and learning. Leaders may need to consider a new type of leadership from the user-design facilitator to help them see how these power shifts are not threatening, but rather healing for their organization.

> What type of leader do you think is best suited to user-design? Why?
>
> What type of leader do you have in your own organization? What impact do you anticipate this might have on any user-design team?
>
> Can you sketch out a basic cost-benefits analysis for your organization's user-design effort?
>
> What do you think will be your own biggest challenge in working with leaders on user-design efforts?

7

User-Design and Systemic Change

INTRODUCTION

To this point, you have built a number of crucial understandings for the implementation of user-design. You've learned what user-design is about, its history, and research. You have explored a number of tools such as scenario planning and design-based research to conduct user-design. You have learned about handling conflict and leadership issues and now we can turn our attention to the broader context of user-design. An understanding of user-design in the absence of true systemic thinking is only half as powerful as it can be when it is more completely linked to the broader contexts that systemic thinking and systemic change force us to consider.

Thus, user-design is most powerful when it is embedded in the broader context of systemic change and improvement (Carr, 1997). In earlier work, I have laid out the distinction between systemic and systematic, making the point that it's more than just two little letters (Carr, 1996). Systemic is best thought of as holistic, contextualized and stakeholder-owned while systematic can be thought of as more as linear, generalizable, and typically top-down or expert driven. These distinctions are useful, but incomplete as a discussion of the whole systemic change context. We have already seen the ways that user-design and systemic change are linked through theorists such as Bela Banathy, Charles Reigeluth, and Patrick Jenlink. But what we need to do now is to really put on a systemic thinking lens so as to make the most of user-design opportunities.

SYSTEMS THEORIES

Systemic change has its foundations in systems theories, both hard systems stemming from engineering roots (e.g., Demming, 1982; Hall, 1962; Hitch, 1955; Jenkins, 1972; Quade, 1963) and soft systems which are less completely under the control of the designer than their engineering counterparts (e.g., Boulding, 1985; Checkland, 1999; Churchman, 1968; Hutchins, 1996). The application of systematic approaches to instruction (e.g., Dick & Carey, 1996; Gagne, Briggs, & Wager; 1992; Seels & Glasgow, 1990) has provided a jumping-off point for considering systems more broadly, such as cultural change and organizational design.

Systems theory and systems design are the foundations for systemic change approaches which are enjoying a huge interest across all fields today. Just recently I was stunned to hear on CNN (the Cable News Network) the term *systemic-change* employed in three different stories about politics, health care and corporate finance within a mere three hour period. Systemic change is certainly a concept that we, as a culture, think we understand. However, systemic change is a process that is based on systems theory and systems design principles which are too often ill-understood by the general population. Systems theory embraces the importance of a global perspective, multiple components, interdependencies, and interconnections in any system (von Bertalanffy, 1968). In addition, the recognition that change in one part of a system necessarily alters the rest of the system is a cornerstone of systems theory (Capra, 1982).

However, this recognition—the idea of ripple effects—in and of itself is not sufficient for an effort to be considered systemic. That is, changing one major system component (piecemeal change) with the understanding that this will cause changes throughout the system does not recognize the inherent nature of systems to avoid change and the impact of interrelationships—it is not a systemic approach to change. As Harman (1984) writes of piecemeal change, "manipulating such obvious variables as budgets, curricula, organizational strategies, etc. would produce the appearance of change, but not much real improvement in outcomes" (p. 3).

Embeddedness is a central concept in systems theory; recognizing that any system-of-interest is embedded in some larger suprasystem and is made up of subsystems. To take the corporation as an example, a business is embedded in a national economic system and a larger global economy. In the same moment, a corporation is made up of many subsystems such as training, personnel, R&D, and production systems. This is important to user-design because the consideration of who the users are is driven by how we define the embeddedness of the system of interest. We can take this same idea to public schools and see that they are made up of subsystems such as curricula, classrooms, communications, leadership, followership and so forth and they are embedded in larger suprasystems such as communities, state departments of education, even the national or global education system if the broadest possible spectrum were considered.

SYSTEMIC DESIGN & CHANGE

As most students of instructional design are already aware, the term *systems design* or a *systematic approach* is frequently used when referring to instructional design processes. Perhaps the most famous and seminal book on the topic of instructional design is Dick and Carey's (1996) text *The Systematic Design of Instruction.* Thus, the systems approach has always been deeply integrated into the psyche of novice and experienced instructional designers. We will explore the connections between user-design and instructional design practices more completely in chapter 9 as a final capstone to all we've learned about user-design. Here, however, it is sufficient to say that the more general class of systems design concerns itself with the process of creating something new by applying a methodology that is grounded in a set of design principles (Jones, 1970; Petroski, 1994; Rowland, 1994; Warfield, 1994). Gardner (1963) describes the fundamental difference between future-oriented design practice and past reform efforts:

> Over the centuries the classic question of social reform has been, How can we cure this or that specifiable ill? Now we must ask another kind of question: How can we design a system that will continuously reform (i.e., renew) itself, beginning with presently specifiable ills and moving on to ills that we cannot now foresee? (p. 5)

This orientation toward design and creation coupled with an understanding of the static tendencies of social bureaucracies and organizations causes many systemic change theorists to call for entirely new systems of human performance (e.g., Banathy, 1991). It is my view that a truly renewable design/continuous change effort can only be a reality when users are engaged in the design process. This process enables users to experience the iterative nature of change and design, thereby building their capacity to accept what might otherwise be seen as an inefficient continuous team process.

When we think systemically, we must see the world BIG. Systemic change is broad in scope and large in scale. It entertains the whole system as the context for understanding change and organizational learning. Skolimowski (1985) writes of the tie between wholeness and community participation:

> ...wholeness means that all parts belong together, and that means they partake in each other. Thus, from the central idea that all is connected, that each is a part of the whole, comes the idea that each participate in the whole. Thus, participation is an implicit aspect of wholeness.

As an example, Tyler Volk (1995) and Gregory Bateson (1979) identify relationships between nature and social organization. In order for us to be parts of whole systems of society and whole systems of ecology, it is necessary for us to actively participate in the making (or unmaking) of our society, democracy, air, and water. Systemic change recognizes the importance of user responsibility within an organization.

Thinking this way used to be unheard of. I recall the first time I taught my systems course here at Penn State to an all-teacher class and was thrilled to see the lights switch on—one by one. There was a great deal of cognitive struggle to understand these fundamental concepts of embeddedness and interconnection.

I remember quite clearly the feeling among my students that the governor of our state needed to take this course so he could begin to think systemically—in fact all our politicians would do well to think in these broader ways. But what has really surprised me is the ways in which our society has come to accept these broad notions without any real changes in behavior. We do recognize now that we are a global economy; we know that the things we do impact the rest of the globe. Yet we continue to consume energy at rates that exponentially outpace developing nations. We ignore the billions of people in China and India insisting on a naïve isolationist notion while paying lip service to the systemic nature of a global society.

The ISSS (2003a) recognized this interconnectedness globally and the potential of unleashing dialogue within a worldwide community:

> Using methods like these, we can generate islands of effective democracy. On a globe where all of us are enmeshed in systemic and chaotic webs, what we do in Dubuque, affects a lecture in Tehran. Information about successful change drives rapid evolution. Successful agoras—in neighborhoods, villages, nations—model healthy behavior for the world. They spread the contagion of hope. (ISSS, 2003, p. 11)

I would caution the reader here to be mindful of what systemic thinking really is and what it is not. There has become a popularized notion of systemic thinking recently and, in my view, this has done more harm than good to the overall movement. No changes in power dynamics

USER-DESIGN AND SYSTEMIC CHANGE 89

have been effected, no significant shifts in mindset can be observed. In fact, I would argue that our world is more entrenched than ever in attempts at piecemeal change. What I am calling you to do is shed this more systematic way of thinking, planning, and even acting and move to broad, holistic, systemic ways of thinking. This shift will be essential to your development as a user-design facilitator. One way to do this is to read the basic systems theorists such as vonBertalanffy, Boulding, and Rapaport. Later in this chapter I will share some basic systems theorists for your review; and of course, you should find alignment with certain theorists who you intend to pursue and read further. Naturally, it would be impossible to express all of their primary learnings within this text, so I instead refer you to some short snippets to read as sources of inspiration for your own development as a systems thinker.

BECOMING A SYSTEMS THINKER

Learning to understand a very different way of seeing the world is not an easy challenge. I truly believe that the first step is to engage some of the significant works in the field of systems theory as I've already suggested. Based on this idea, I'm offering some initial thoughts on systems for you to read and review and begin to think about. Explore these within their context if you can, but try hard to let the ideas wash over you. Systems is not about understanding pieces, and I've found that trying to understand systems theory or to think systemically by unpacking every word in the systems theorists' works is self-defeating. Instead look at these quotes from three systems theorists and my commentary on them as a beginning step into the world of systems thinking and begin to see if it changes your perspective, your cognition, your ideas, the way you see things.

vonBertalanffy

Often thought of as the father of general systems theory, vonBertalanffy was a biologist who sought to find threads in order to link disparate fields to better serve mankind.

> Our civilization seems to be suffering a second curse of Babel: Just as the human race builds a tower of knowledge that reaches to the heavens, we are stricken by a malady in which we find ourselves attempting to communicate with each other in countless tongues of scientific specialization.... The only goal of science appeared to be analytical, e.g., the splitting up of reality into ever smaller units and the isolation of individual causal trains....

> We may state as characteristic of modern science that this scheme of isolable units acting in one-way causality has proven to be insufficient. Hence the appearance, in all fields of science, of notions like wholeness, holistic, organismic, gestalt, etc., which all signify that, in the last resort, we must think in terms of systems of elements in mutual interaction." Davidson, 1983, p. 184; quoting vonBertalanffy

There isn't a great deal of confusion for me in this quote. It seems pretty clear to me that he is criticizing the increasing levels of specialization and atomization among all fields of study and is encouraging us to talk to one another—really talk to one another. This may mean that we will have to learn one anothers' languages; that we will have to set aside our own desire for power to impress upon others our own language and our own ways of knowing; that we will have to really want to change the world—together.

Checkland

For me, Checkland is the father of the other systems theory—the soft side of systems theory. And this is the most important type of systems theory to me as it helps me to understand the nature of social systems which I am most interested in. Now one important thing here that cannot be overlooked or ignored—this is not to imply that there is a typology of systems that can cause greater disintegration among systems theorists themselves. That is, when will the systems theorists so completely shroud their own field in jargon and constructs such that they cannot even talk to one another?—a smaller tower of Babel, vonBertalanffy might suggest. Soft systems, though, was the first major break from the hard sciences which recognized that social systems do not necessarily behave in the same fashion as hard scientific systems and therefore may need to be approached in somewhat different fashion.

> The central concept 'system' embodies the idea of a set of elements connected together to form a whole, this showing properties which are properties of the whole, rather than properties of its component parts. (The taste of water, for example, is a property of the substance water, not of the hydrogen and oxygen that combine to form it.) The phrase "systems thinking" implies thinking about the world outside ourselves, and doing so by means of the concept 'system.' (p. 3, Checkland, 1999)

I really love this water example of course, we don't understand water as a concept through understanding its components at the elemental level. Rather we know "water." A corrallary to this is also that when a thing is broken into its parts, we can see a reflection of the entire whole

USER-DESIGN AND SYSTEMIC CHANGE

in its parts. This is the idea of *fractals*, a physics concept which is highly related to the idea that a system reflects what its whole is rather than its parts. We can see fractals, for example, in sea coral where when we break off a small part, it reflects the structure of the larger coral—as a crude example. What I like best about this quote, though, is his assertion that we must think about the world outside of ourselves in order to really call ourselves systems thinkers, and that we do that by understanding what system means:

> The three problems for science considered above—complexity in general, the extension of science to cover social phenomena, and the application of scientific methodology in real-world situations—have not yet been satisfactorily solved, although some progress has been made. Had they been solved, it is unlikely that systems thinking and the systems movement which unites systems thinkers in many different fields, would exist in their present form. This is not to say that systems thinking has made spectacular progress—indeed, on the whole the substantive results from the systems movement are still meager—but the existence of the movement at all is a response to the inability of reductionist science to cope with various forms of complexity. Systems thinking is an attempt, within the broad sweep of science, to retain much of that tradition but to supplement it by tackling the problem of irreducible complexity via a form of thinking based on wholes and their properties, which compliments scientific reductionism. (p. 74, Checkland, 1999)

I am refreshed, here, by Checkland's admission that the systems field still leaves a good bit to be desired in terms of the advances that systems thinking and systems theories have made. But, I do agree with his assertion that the presence of systems thinking means that we needed something else—something in response to a set of sciences that were not doing what we needed them to do. Our attempts to take pieces and parts out of the whole and then try to fix them in an attempt to fix the whole simply hasn't worked effectively. He is, in my opinion, generous to say that seeing wholes and systems thinking are complementary to the attempts that scientific reductionism has made at solving our big world problems.

Banathy

> The ontological task is the formation of a systems view of what is, in the broadest sense a systems view of the world. This can lead to a new orientation for scientific inquiry. There are two great philosophical alternatives of the intellectual picture we have of the world. One view is that the world essentially consists of things. The other view is that the world consists of processes, and the things are only "stills" out of the moving picture. Systems philosophy

> developed as the main rival of the "thing view." It recognizes that primacy of organizing relationship processes between entities (of systems) from which emerge the novel properties of systems. (Banathy, 2003, p. 3)

Bela Banathy's work is among the most profound in the field of educational systems deign. And he is largely responsible for this very book you are reading since his work laid the foundation for user-design within the field of Instructional Systems Technologies more generally. Sometimes I find Bela's work difficult to understand, particularly if I pull it apart into its pieces. But when I read Bela's work for its spirit and wholeness, I come away satisfied. The major issues I think that Bela is bringing out here in these passages are the importance of approaching things from processes rather than pieces or "things" as he calls them. There is some trouble with his characterization of only two great movements in the world this sort of dichotomous thinking can lead one down an improper path, so I would tend to shy away from taking that part of the message too seriously, but the basic idea that the thing view is an outmoded way of thinking and that we should move more to processes is an easy bit to understand. Making it work in your life may be another matter:

> Epistemology deals with the general questions of how do we know what we know, how do we know what kind of world we live in and what kind of organisms we are, and what sort of thing the mind is. The ancient questions of whether the mind is immanent or transcendent can be answered in favor of immanence. Furthermore, any on-going ensemble (system) that has the appropriate complexity of causal and energy relationships: (a) will show mutual characteristics (b) will compare and respond to differences (c) will process information (d) will be self-corrective and (e) no part of an internally interactive system can exercise unilateral control over other parts of the system. The most significant guiding principle of systems inquiry is that of giving prominence to synthesis; not only as the culminating activity of the inquiry (following analysis), but as a point of departure. This approach to the "how do we know" contrasts with the epistemology of traditional science that is almost exclusively analytical. (Banathy, 2003, p. 344)

Now all this talk about epistemology and ontology and so forth tends to be difficult philosophical language to untangle. And I haven't found that much to be gained by the careful untangling of it, though the conversations with philosophers can be very interesting indeed. The most important thing about this particular passage, to me, is the nature of systems is being more fully revealed here. We can begin, then, to understand that systems are interdependent, they relate to one another, and

USER-DESIGN AND SYSTEMIC CHANGE 93

they will engage in processes, so that old input-process-output model we see in the hard systems or computer systems world is probably not entirely out of the question from Banathy's vantage point. I like this idea that they will be self-corrective; to me, this implies a certain amount of inertia that then has to be overcome if a system is to be changed dramatically; self corrective may not always be in a positive direction at all. Self-corrective may imply self-interested, thus communicating a certain sense that systems will maintain themselves as they are unless a serious threat occurs and they must change to survive. Naturally, Banathy's point regarding no single part of a system exercising control over another is leading toward a user-design understanding: the system must design and control itself. Again, the primary message in both of these Banathy passages is that understanding a system is not about tearing it to pieces to understand each bit conceptually, but rather to synthesize and to take synthesis not as an end point, but as a starting point:

> By observing various types of systems and studying their behavior, we can recognize characteristics that are common to all systems. Once we have identified and described a set of concepts that are common to the systems and observed and discovered among some of them certain relationships, we can construct from them GENERAL SYSTEMS PRINCIPLES. Thus a system principle emerges from an interaction/integration of related concepts. Next, we are in the position to look for relationships among principles and organize related principles into certain conceptual schemes we call SYSTEMS MODELS. This process of starting from observation and arriving at the construction of systems models constitutes the FIRST STAGE of developing a systems view. (p. 15, ISSS, 2003b)

> Systems thinking is a property of the thinker, who organizes internalized systems ideas, systems concepts, and principles into an internally consistent arrangement, using a systems way of viewing and understanding, in order to establish a frame of thinking. As we observe what is "out there,' this frame of thinking enables us to reflect upon what we experience; thus we construct our own meaning. We create out own cognitive map, which is our own interpretation of the out there. As we view and work with social systems, systems thinking enables us to create our own cognitive map of the systems of our interest. Banathy, 1996 p. 156

> As we speak of the design of an educational system that carries out societal functions, we can no longer use the term 'system' loosely as we do in our daily lives. (Everything is a system today, thus the meaning of the term is trivialized.) Systems thinking will provide us with new insights into what the real meaning of a system is. It will empower us to approach educational reform with a new "mindset" from the perspective of a systems view of the world. (p. 31, Banathy, 1991)

AN EXAMPLE OF USER-DESIGN AND SYSTEMIC CHANGE

An example may help to illustrate the kind of thinking I'm challenging you to really engage. Let's look at the development of the Internet as an instantiation of systemic thinking and user-design. One of the more interesting texts in this regard is *Linked* (Barabasi, 2003). The Internet may have been the ultimate in user-design experimental spaces. Here, initially, was the creation of so much content by individual users of the Internet. In the beginning, the Internet was truly an open source technology anyone could come and create whatever they wanted—truly without censor—no matter how roughly constructed. The tools became increasingly available and accessible to all users. There was early hope that the Internet may be just the postmodern tool to overturn authority and expertism and open the flow of all information in a democratic fashion. And every once in a while a shining example or opportunity to renew the democratic nature of the Internet crops up such as the current Wiki movement to build open source encyclopedic entries and dialogue. But, if you think systemically and understand all the various powerful forces at play, you can see why this potential user-design experiment did not actually sustain the open nature for the long haul. Hirschkop (1998) suggests that it would have been "poetic justice" for corporate power to sell its "gravedigger to the masses in inexpensive, high-tech form" (p. 217). And McChesney (1998) writes of the same issue:

> The Internet has opened up very important space for progressive and democratic communication, especially for activists hamstrug by traditional commercial media. This alone has made the Internet an extremely positive development. Some have argued that the Internet will eventually break up the vise-like grip of the global media monopoly and provide the basis for a golden age of free, uncensored, democratic communication. (p. 21)

However, this reading of the potential of the Internet ignores the warnings issued to us by the postmodernists and critical theorists. We must ask ourselves questions such as, who will benefit from this technology?, and who will be left behind? We must take a systemic view of the whole society to understand that if this media is going to be powerful, it will almost certainly be wrested from the hands of the users and turned over, instead, to experts and authorities. Noam Chomsky (1998) makes this point clearly:

> It's particularly harmful to democracy when media systems are in the hands of private tyrannies .... Here's this huge system, built at public expense. Most media analysts with their heads screwed on see, and even report, that

USER-DESIGN AND SYSTEMIC CHANGE 95

> it's very likely going to end up in the hands of a half-dozen megacorporations internationally. (p. 188)

So, we need to take a broad systems view to understand the impact of the internet as a system within the larger system. And McChesney (1998) points out that this system exists within a system of rapid commercialization—some would say rabid capitalism:

> Yet whether one can extrapolate from activist use of the Internet to seeing the Internet become a democratic medium for society writ large is another matter. The notion that the Internet will permit humanity to leapfrog over capitalism and corporate communication is in sharp contrast to the present rapid commercialization of the Internet. (p. 21).

Thus the systems perspective helps us to begin to ask the right questions and analyze the systemic/cultural context sufficiently to imagine predictable failures (as Sarason (1998), calls them). Hirschkop (1998) clearly states that the availability of information on the Internet alone will not alter the capitalist political structure of power currently in place.

> It was not a momentary loss of judgment (or the need for a savior in a dark age) that led so many on the left to delude themselves about the democratic possibilities of the new media. Belief in what technology could do depended on a fundamental misreading of the nature of liberal-capitalist states and of political power in the late twentieth century. (p. 215)

THE STRUGGLE

Given this history and critique of the development of the Internet, what might user-design have looked like if it had been more effectively utilized? And how could a systemic lens have helped us? If we had actually applied systems perspectives and user-design to this particular example, the result would look like struggle. Lofty goals such as equitable educational opportunities, the elimination of racism, or classism, or true democracy are not easily attained goals that arrive through the trojan horse of technology. Indeed, it is only through true struggle that we might ever see the realization of the underlying goals of critical theory, systems theory and user-design. As Hirshkop (1998) eloquently writes:

> When the very structure of a society depends upon a lack of democracy, however, democracy will depend upon a fight, and upon social forces with the interests, will, and intelligence to struggle for it. Technology will doubtless have a role in this struggle, but it offers no shortcuts: one cannot buy

democracy off a shelf, or download it from a Website. It demands courage, fortitude, and political organization, and, as far as we can see, Microsoft has yet to design software that can deliver these. (p. 217)

In fact, Microsoft is the opposite of a democratic organization. I was very surprised to discover Microsoft conducting push advertising within my new computer recently purchased here at the university. When I first powered up my Mac, I used the transfer process they so wisely provided for me and moved things from my old to my new machine so that the transition has been largely seamless from one machine to the next. However, Bill Gates had other ideas. The Office "Test Drive" was installed on my machine and was pretty well hidden within the new software. But when I went to open my documents, spreadsheets, or databases, I was given 30 days or 27 days, or so forth to purchase the new version of Office or, well, or else, I guess. I managed to delete the darned thing, but this dialogue box warning me that I needed to "Buy Now" popped up pretty much every time I wanted to save, and what was worse, the program made it impossible for me to print any of my Office documents at all until it was uninstalled. Pretty democratic. I fear Mr. Gates would very much like to turn us all into Microserfs.

Struggle is going to be a part of the user-design approach. To pretend otherwise is to mislead. That's why there is such clear discussion in this book on conflict, and on when it is wise *NOT* to employ a user-design approach. In this culture, it is not a given that this struggle can even be entertained within the current power paradigms. But we have to try we have to struggle; to fight the good fight. When we understand all of the possibilities and potentials of truly engaging the users in their own creation of human learning systems, how can we ignore that power and potential? But in order to make the struggle worth the while, however, we have to understand the world systemically and we have to employ good user-design tools that take best advantage of the systemic change context.

Understanding systemic change and user-design means that you cannot come into a setting with a ready-made set of answers. True systems thinkers realize that any solution imposed on a group of people will likely fail for many systemic reasons. When I am asked what I think new systems of human learning should be like, I'll sometimes point to some colleagues' work and suggest careful consideration of highly divergent opinions. This reluctance is not an expert withholding the "light." Much the opposite, the only light I have to shed is that which allows others to create their own enlightenment. David Purpel (1989) sees the expectation of communities to find light bringers as absurd:

USER-DESIGN AND SYSTEMIC CHANGE

> It is surely not for me to provide the last word on what the broad educational framework ought to be, never mind the absurdity of an expectation that I or anyone else could also provide a detailed blueprint of implementation and practice. (p. 279)

Each user-design team must be treated as its own unique culture, within its own unique context. It is very difficult to find very much to generalize across teams, in fact. This sort of generalization is precisely what's gotten us where we are and perpetuates the dominant Cartesian model of the world as a machine. Relying on expert opinion generalizing across very different contexts washing out the diversity of viewpoints, desires and ideas trapped inside indigenous knowledge and offering (sometimes forcing) visions for adoption on the users does not meet the directives of true user-design, the ethics of systems theory, nor the standards of critical theories. For instructional designers, we have always understood, to a certain extent, that our job is about changing schools and organizations. It's not mere improvement—our goals are loftier than that. We know that instructional design can unlock the potential of human performance within organizations and radically alter schools if it is really given a chance. But that chance may come more through release of the ideas than control of them. In just the same way that we now can easily accept that a one-size-fits-all school system is not viable—that there is no one answer product-wise—perhaps we can now begin to accept that there is no one single answer process-wise either.

This is not to say that the traditional models of instructional design have no place. Indeed, this text is meant as a companion piece to the traditional texts in the field. Rather, we have to understand user-design in instructional design as a way to leverage systemic lenses for broad change and system-wide transformation.

CHAPTER CONCLUSION AND THOUGHT QUESTIONS

In this chapter I have challenged you to put on a different lens to see the world. I want you to look through a set of systemic glasses and see the impacts of changes on entire systems, the embeddednes, interconnectedness, and ripple effects. The best way to do this, I have argued, is to spend time with the original systems writings. A word of advice here may be worthwhile. My experience of systems literature has been that it's not easy to read, and that I actually get most out of it by simply allowing the ideas to wash over me, rather than dissecting them in a more Cartesian fashion. Taking the literature holistically may be a first step to learning to think and live holistically as well. We looked at the development of the Internet as an example of a failed user-design systemic change and again I challenge you to think of some other examples in your own life or organization which may help you to look at the world systemically and ask the right questions so you can see predictable failures before that path is trod. Understanding the struggles and the importance of leaving pre-made solutions behind concluded this chapter. Ultimately, seeing the world in this way will elevate your thinking beyond minutia.

I have a colleague who insists that you can think systemically but that you can only act systematically. He gives as an example going to the grocery store and using a list to guide your efforts and be more efficient. Maybe he's right, but I suspect that really seeing the world systemically will lead you to a life of opportunities to act systemically as well. David Bohm seems to agree:

> Indeed to some extent it has always been necessary and proper for man, in his thinking, to divide things up, if we tried to deal with the whole of reality at once, we would be swamped. However, when this mode of thought is applied more broadly to man's notion of himself and the whole world in which he lives (i.e., in his world-view) then man ceases to regard the resultant divisions as merely useful or convenient and begins to see and experience himself and this world as actually constituted of separately existing fragments. What is needed is a relativistic theory, to give up altogether the notion that the world is constituted of basic objects or building blocks. Rather one has to view the world in terms of universal flux of events and processes. (David Bohm, 1980)

What does it mean to see the world systemically?

Who are three systems theorists whose work you intend to explore further?

What is the relation between systemic thinking, systemic change, and UD?

How is the Internet an example of UD? How is it not?

In what ways do you believe that systemic thinking will help you in your UD efforts?

What does it mean to say that UD is embedded in a systemic change context?

8

User-Design and Performance Technology

We see the world from a systems lens, and we understand the relationship of user-design to systemic change. But, this language is far more common in public schools and even in higher education than it is in corporate or non-profit organizations. Instructional design, of course, happens in many contexts from museums to jet-fighter pilot training bases. And, so it is important here to distinguish what is meant by user-design in organizations which do not focus so much on systemic change for their innovative energies, but rather use the term *Performance Technology* (PT). In this chapter I make the link between PT and instructional design and user-design as constructs and discuss the ways in which each informs the others. It is important to understand that PT is really a special case of systemic thinking and so the case for this connection is made early in the chapter. This does not mean that PT can be truly merged with user-design since their projects are entirely different. PT maintains, as we'll see, a focus on users but to the benefit of the organization's bottom line from a leadership perspective or a stockholder perspective rather than from a user's perspective. Thus the critical power shift is largely missing from the PT paradigm to this point. Finally distinguishing between the ways that the PT community uses stakeholder approaches from the true user-design effort helps you to understand the challenges associated with user-design that are unique within the corporate context.

PERFORMANCE TECHNOLOGY AS A SPECIAL CASE OF SYSTEMIC THINKING

The basic ideas of systems thinking have been laid out in Chapter 7 as embeddedness, interconnectedness, interdependency and stakeholder participation. How do these ideas play in the corporate setting? Actually, it is relatively simple to see that the corporate language of translates very closely into the systemic change context. A number of recent texts on performance technology have highlighted the importance of systems thinking, and Senge's landmark text on *The Fifth Discipline* (Senge, 1990) (which is systems thinking) has been translated into corporate change models to great advantage. Bell and Forbes (1997) explored the importance of systems thinking within performance technology and go so far as to advocate for more complex understandings of even chaos theories:

> The generalization of the feedback process is the systems thinking now popularized by Senge (1990). Even the simplest systems soon fail when treated as static. Wheatley (1992) takes the metaphor further and links organizations with quantum mechanics and chaos. Waldrop (1992) seeks order in chaos more broadly, showing that humans can never be "outside" a system they influence (can never ultimately "win" if someone or something "loses") and finds innovation at "the edge of chaos." Patterns and systems self-organize. This kind of thinking is increasingly a requirement in managing performance. (p. 378)

Clearly, the understandings inherent in systems thinking are seen by at least some in the organizational change field as critical to success. Ford (1999) agrees with this position and also emphasizes the importance of stakeholders in his look at performance technology in organizational design:

> Systems theory pervades all of the theory and practice of OD. It involves the assumption that all parts of the organization are connected, related, and interdependent. An organizational system can be characterized as a continual cycle of input, transformation, output, and feedback, whereby one element of experience influences the next. ... In addressing an organizational issue with a systems perspective, one maintains a focus on stakeholders—those who have interests relevant to that issue. Such a perspective attempts to understand the points of view of each interested party. In particular, one must understand and acknowledge that there is a possibility of both conflicting as well as common interests among stakeholders. Thus the systems framework forces one to consider to what extent and in what ways the interests of the various stakeholders are involved in decision making and in what ways attempts are made to foster common rather than competing interests. (p. 253)

Thus, the translation from the language of systemic change into the language of performance technology is not so far from base. And clearly, the performance technology theorists have recognized these same forces of power and influence on the decision-making process. However, in general, the literature in performance technology tends *not* to take a systems perspective, and even less often takes any kind of a stakeholder participation stance. The vast majority of literature in performance technology is focused on sharing a variety of particular interventions such as job redesign, incentive systems, or strategic alignment. The interventions range from narrow to broad, and the broader the intervention the more likely it is to embrace systems thinking, and/or stakeholder participation.

PERFORMANCE TECHNOLOGY AND USER-DESIGN

In the relatively rare occasions where systems thinking and stakeholder participation are important in the PT literature, the approach ranges from the more common superficial accounting for stakeholder opinions through something that comes closer to true empowerment. One of the important distinctions among the approaches is the extent to which the authors express the value of a stakeholder approach. Rothwell (2005) addresses the necessity of consulting with stakeholders by asking the following set of questions

> Who has the biggest stake in the performance gap? Why? What results do stakeholders most want to achieve now and in the future? When are the stakeholders most interested in finding out about the performance gaps? Now or in the future? Why? What evidence will be most persuasive to those who must decide whether to take action? What approaches to presenting evidence will the chief stakeholders find most persuasive as they decide whether to act? How willing are the stakeholders to participate in troubleshooting a performance problem or seizing a performance enhancement opportunity? Why? (p. 142)

The set of questions that Rothwell offers us clearly is founded on the somewhat colonial idea which the stakeholders must be convinced or persuaded that the solution that has been determined based on the identified performance gaps and prescribed by experts in training design and development is what is best for them. In a very similar vein, Schein, (1999) outlines the elements of change programs which aid in cultural innovation by alleviating resistance. He suggests a clear vision, corresponding structural changes, redesign of organizational systems, education and training programs to learn new behaviors, coaching and employee involvement. Schein, then, in some ways comes closest to user-design in his final two elements:

> *Coaching and other informal learning opportunities:* Informal learning helps employees to internalize the new behavior, values, and vision. *Employee*

> *participation in the design of the change and learning process:* Those whose behavior, values, and assumptions must change need to participate in designing the change and learning processes. Their participation lessens their anxiety and promotes their acceptance of the changes. (p. 130)

Here, then, is a clear recognition of the value of user participation in the design process. And this recommendation is predicated on the value that UD manifests in adoption rates. Unfortunately, this doesn't go far enough to acknowledge the value of improved designs. The resulting products and processes are better for the organization. Merely seeing employee participation as a way to grease the inevitable slow skids of adoption ignores the benefits to the organization by focusing only on getting users to do what others' have created as solutions. Ford (1999), however, discusses the process of taking a systems perspective in a chapter on Organizational Development in the *Intervention Resource Guide* and recognizes the potential of improved solutions. This discussion of how to best involve people in the change process even recognizes a certain sharing of expert power—a rare recognition in the field of PT:

> ... many OD change efforts are based on the assumption that those who will be affected by the change need to be involved in that change. Without this involvement, most change efforts are doomed to failure. The kind of participation and involvement that is required extends beyond management. Research and practice have shown that increased involvement is desired by most people, has the ability to energize performance, produces better solutions to problems, and greatly enhances the acceptance of decisions. Some rules of thumb for involvement include the following: Involve everyone who is part of the problem or part of the solution. Treat those closest to the problem as experts on the subject. Provide the support and power needed by those closest to the problem in order to make decisions. (p. 254)

It is important in PT literature to understand how to unlock the potential of human performance and so, in most cases there is a focus on how the organization can meet the needs of the employee. For example, Bell and Forbes (1997) outline the results of Kovach's (1987) empirical research on what workers want as a way to raise productivity. Some of these findings are useful in thinking about the added benefits of user-design from within performance-oriented organizations. For example, the number one reward employees seek is interesting work and the third most cited was "a feeling of being in on things" (Kovach, 1987, 154). User-design can comfortably accommodate both of these rewards by making work more interesting and engaging as well as helping to engender a feeling of being in on things. The very process of engaging users in the creation of their own systems builds ownership, but also an interest in their own work processes and a sense that they are included and valued.

USER-DESIGN AND PERFORMANCE TECHNOLOGY

Within the field of performance technology there are even some orientations that at least begin to honor learners as users. Wydra and McKenna (1999) have a chapter in the *Intervention Resource Guide* on "Learner Controlled Instruction," which focused on allowing for certain learner controls within constrained instructional parameters for learners within their own learning environments. The suggestion is that they control things like pace or learning site, but not the goals or objectives. It does not really encourage any engagement of learners in the process of instructional design, but rather suggests ways to take instruction into a space of consideration for learner desires. This is at least a beginning within performance technology toward user-design for instructional design. But, it unfortunately does not go far enough toward creating optimal user-design spaces within the performance-oriented organization.

GUIDANCE FOR PERFORMANCE TECHNOLOGISTS

Performance Technologists wishing to employ user-design have particular constraints which must be considered prior to embracing a very different sort of model from those that dominate performance-oriented organizations of all types.

PT and ID

Despite the shining moments outlined above in which PT literature recognizes and honors users and learners, in general, performance technology-oriented organizations are heavy on singular leadership. The ideas are much more oriented toward control and bottom line production than perhaps the earlier portion of this chapter might lead you to believe.

An example is the very recent book by Michael Dreikorn (2004), *The Synergy of One*. In this book, categorized as PT, Dreikorn posits a model of leadership: Integrated Performance Leadership. The model is highly unidimensional, focusing on leadership as the force for change.

> A single leader should lead the deployment efforts, and a cross-functional team should be created to drive the work of change. The process of change should include the entire value stream, and engagement should be measured. Those who opt out of the implementation process need to be bought back into the fold or asked to find other employment .... Set the course, create the team, educate the people, develop the plans, measure and celebrate success, keep people engaged, stay the course and make it happen! (p. 126)

Pretty clearly, there is one player here, one expert, and the values of Schein (1999) and Ford (1999) are not likely to permeate the organization headed by an Integrated Performance Leader. So how does someone within a traditional training and development department make the first steps toward a user-design project?

First, we need to return to the questions that were asked in the first chapter of this book in terms of whether or not user-design really makes sense. You may recall that questions included whether or not the leadership is secure and therefore willing to empower users, whether user resistance is high, the problem is systemic, and better products and processes are desired. The performance-oriented organization may have some of these hallmarks, and it is likely in those cases that user-design will be more successful.

Negotiating with leaders in the case of corporate settings is very important. Remember that leaders of corporations tend to have come from pressure cookers of performance expectations. They are not likely to be open to long timelines for user-design, and probably this will be the hardest thing about creating an effective user-design project in a for-profit corporation. Where schools are used to relatively long timelines (although with the current climate of accountability this may change in the near future), corporations are simply not comfortable with the inefficiencies of user-design. Despite many of the most compelling advantages including better processes and products, unless the organization is either in extreme need of new ideas or is *very* secure in its position in the marketplace and willing to experiment, the leader may not be game for user-design.

As a user-design facilitator, it is essential that you not try to "sell" management of performance-oriented organizations unrealistically on the benefits of user-design. The very best policy is true, basic, honest disclosure. Any leader must know precisely what they are in for. The costs are high, the benefits are high too and a cost benefits analysis can show leaders the benefits; but be cautious not to undersell the costs.

Because corporations exist in much more pressurized environments, consistently looking for the next threat to their position in the market, they are going to exhibit more conflict and, in some ways, more dramatic conflict than you may find in non-profits or schools. Therefore it is important to consider, in advance, ways to deal with conflict in productive ways. Companies, like schools, and even more so like families, tend to want to cover up conflict, to avoid it at all costs. Refer back to chapter 5 for a refresher on conflict management in user-design.

Finally, user-design in performance technology-oriented organizations must be seen as a special case of systemic change. Zemke (1989) points to the systems approach in performance technology:

> Perhaps the most encompassing of models is the acclaimed systems approach—usually referred to by its advocates as "The Systems Approach" or the Instructional Systems Development (ISD) model … . (p. 19)

> Overall, it appears that something we can call a systematic approach to the design, development and delivery of training is alive and well in the field.... But despite the jumble and variety of approaches we found, one thing is clear: Those who practice a systematic approach fare better in their organizations than those who don't. (p. 22)

But, performance technology considers the systems approach as systematic as much as systemic; within user-design we are more interested in the systemic thinking than the systematic acting. Corporations, however, are often on the opposite side of this, being more concerned with systematic action than systemic thinking, which, for the most part, is still a foreign language. Thus, understanding user-design within the performance technology organization requires a certain illumination of the systemic nature of performance technology itself among leaders. Showing leaders the ways in which performance technology is clearly a systemic thinking technology can help to go a long way toward embedding the user-design initiative in a systemic context, which is where it is most powerful.

CHAPTER CONCLUSION AND THOUGHT QUESTIONS

Performance technology offers a unique perspective on systemic change. As a set of interventions that enhance the ability of organizations to improve the human performance within primarily corporations, performance technology can mix with user-design as long as certain cautions are considered. User-design can be seen as out of alignment with the needs of high performance organizations, but in reality, the user-design approach matches well with secure organizations wishing to empower employees and generate significant advances through an influx of new ideas and rapid adoption rates. This chapter has tried to point toward hopeful literature in the performance technology area that accords importance to stakeholder participation, but admits that the current scholarship in performance technology does not go far enough in terms of supporting a user-design approach. Careful attention to leadership, conflict, and systemic issues will ensure a more successful user-design experience.

> How is PT a special case of systemic thinking?
> Why is UD difficult in high-performance organizations?
> What do you believe would be the most difficult aspect of a UD approach in any organization devoted to PT?
> What special considerations should be taken to ensure a positive UD experience in a high-performance work environment?

9

Linking User-Design to Traditional Instructional Systems Design Models

You've come through a great deal of learning to get to this point. And now we can begin to synthesize the ideas presented in this book into the traditional ID models that guide most of us in the creation of human learning environments. To this point you may be feeling that some of the ideas in this book have been idealistic, naïve, or even abstract. You may be trying to understand how user-design leads to the creation of instructional materials. While user-design may be used to create everything from entire organizations to very specific computer interfaces, it can clearly be used to create systems of human learning and, more specifically, instructional moments.

Many of the ideas of user-design are heavily rooted in theories of leadership, conflict, and design and systems thinking. Now is the time to take all of those great ideas and apply them to the creation of instruction. You may be feeling that the ideas presented don't quite "fit" with what you might understand as traditional ID models. This chapter will help you to see the connections and interconnections between these two constructs so that you can put the ideas in this book into immediate use, whether as a full-blown user-design effort, or just a little user empowerment within more traditional ID processes.

One important caveat is necessary here, however it is important to understand clearly that UD and ID come from very different worlds. Their underlying epistemologies or belief systems about how we understand the world are not exactly diametrically opposed, but they are not precisely in alignment either. The designers' world view is perhaps the main difference between these two approaches to the creation of instruction. As such, it is immutable and it is important that this attempt to meld the two approaches does not rob both of their strengths.

First, this chapter addresses the ways that user-design and instructional design are joined by their foundations in design. This leads to a careful explication of the similarities and differences between user-design and instructional design. This comparison in terms of the components of user-design and instructional design naturally extends into a comparison across the two processes so that you can build an understanding of the stages and how they can fit together. A brief fictional description of a typical day in the life of a traditional ID practitioner is then compared with a similar description of a user-design ID practitioner as a way to make the connections concrete.

DESIGN

Both ID and user-design are, by their very nature, design disciplines. They are firmly planted in the world of creation. Harold Nelson said in a recent (2005) interview that, "Design is the means by which humans continue to participate in the ongoing genesis of the real world." That qualifies a whole lot of disciplines as design—from architecture to zoological program development. There is creation and genesis happening all around us, but design is a particular kind of handiwork. Design is intentional and change oriented as Jones points out:

> If we seek a firmer basis for our thoughts we have better look outside the process itself and try to define designing by its results. A simple way of doing this is to look at the end of the chain of events that begins with the sponsor's wish and moves through the actions of designers, manufacturers, distributors and consumers to the ultimate effects of a newly designed thing upon the world at large. All one can say with certainty is that society, or the world, is not the same as it was before the new design appeared. The new design has, if successful, changed the situation in just the way the that the sponsor hoped it would. If the design is unsuccessful (which in many cases is more likely) the final effect may be far from the sponsor's hoopes and the designer's predictions but it is still a change of one kind or another. In either case we can conclude that the effect of designing is to initiate change in man-made things. (Jones, 1970, p. 4)

Nelson's interview (2005) clearly puts design into a category that is in alignment with user-design by identifying the role that service plays in the design activity:

We differentiate design from art and science by the concept of service. Design is defined as service on behalf of someone else—a contractual relationship. Artists and scientists engage in forms of service legitimately focused more on their own interests. Artists express their emotions and feelings; scientists express their curiosity about the world. Designers, however, serve the needs and desires of others. This does not mean that designers are not aesthetic or rational they are both but most importantly, they are empathic .... If designers are going to design learning and teaching systems, they need to be educated as designers and gain experience as designers. They need to know the distinctions between the different traditions of inquiry that determine what the different purposes of learning and teaching systems need to be. Design competence is essential for engaging in authentic design activity.

How Are Instructional Design and User-Design Same/Different From One Another?

Other than their roots within the design discipline, there are a number of similarities and differences across user-design and instructional design (see Table 6). First, in addition to being interested in the creation of human learning environments, both are being applied to education and instruction. In the case of ID, the application is, admittedly, a bit more narrow than the hopes of the user-design movement. Instructional design is, by definition most interested in instruction. Some have asserted that those who are interested in broader issues such as educational systems design (ESD) (e.g., Banathy, Reigeluth, Jenlink) should not call themselves instructional designers and should not be a part of the field because their assertions are less rooted in hard science (Merill, 1996). The user-design process has definitely been interested in broader issues of design having sprung from the ESD movement.

Both ID and UD are impacted by the current advances in learning theories. Because both ID and UD are interested in instruction and education, deep understandings of how people learn and the most recent empirical findings about learning are of great interest to both fields. Both groups have an interest, for example, in the learning sciences and the ways in which technologies (both hard and soft) can be employed to their best advantage in classrooms and organizations. Recent advances in brain research are likely to be more interesting to those in the ID field, while those in the UD discipline may find more application for the results of human-computer interface literatures and findings in stakeholder participation. However, there are some common areas of theoretical interest between the two fields.

TABLE 6
ID/UD Comparison

ID	UD
Design discipline	Design discipline
Application to instruction	Application to instruction and broader issues of educational systems design or even HCI, organizational design, and culture change
Affected by current learning theory advances	Affected by current learning theory advances as well as emerging dynamic systems theories such as chaos and complexity
Highly goal oriented most often based on behavioral objectives	Highly process oriented most often aimed at user empowerment
Systematic approach to align goals, strategies, materials, and assessment	Systemic approach to engage users in creation activities
Founded as a way to advance the application of technologies	Founded as a way to serve others and to empower users through design
Carried out in military, corporate, public school, higher education, and nonprofit organizations	Carried out in schools, higher education, and corporations (not yet in military)
Struggles with issues of identity and effectiveness as a field	Struggles with initial stages of field definition
Organized by specific, usually discrete, and linear steps	Organized in looser phases

ID and UD are almost diametrically opposed as far as their primary orientation is concerned. ID is highly goal-oriented. In the PT world, ID is expected to produce very specific, behavioral outcomes that translate into performance improvement for the organization. Training organizations have become adept at showing their behavioral gains and making sure that their analyses identify the real problem so that their goals align with the needs of the organization. To a lesser degree, ID in schools also is very goal-oriented. While ID in schools tends to skip the needs assessment phase (due to the highly political nature of curriculum setting in schools), their reliance on behavioral objectives is no less profound. User-design, on the other hand, is highly process oriented. This is not to say that those practicing user-design in organizations of all sorts are not concerned with showing the fruits of their labors. Indeed, as this book has illustrated, it is imperative that user designers conduct careful cost–benefit analyses to highlight the impacts of their processes and resulting higher quality products. And even more, it is essential that we not limit ourselves to cost–benefit analyses only. While ID has had a tendency to focus on efficiency and therefore cost–benefit analyses almost to the detriment of other important measures of effectiveness, the user-design paradigm definitely does not limit itself in this way and we should try to capitalize on this opportunity to broaden our consideration of analyses of impact. Thus, the primary orientation within the user-design field is focused on the process of engaging and empowering users in the creation of their own learning. It is, in fact, through this engagement that user-designers find the benefits are most substantial.

As discussed earlier in this book, the UD approach is more concerned with systemic than systematic approaches. ID remains a systematic enterprise. While many understand ID to be a systems approach, this is primarily aimed at aligning the components of the system such as goals, strategies, materials, and assessment into a system which is consistent across stages. In addition, the ID process is rooted in more hard systems constructs such as input, process, and output and iterative processes as metaphors for the systems approach. User-design, on the other hand, is more interested in the systemic approach and the goals are more oriented toward the use of systems thinking for the purpose of engaging users in the creation of their own learning systems. This difference can be confusing to novices in either field because they see a similar language around systems and believe the approaches to treat the term in the same way, which is not really the case. The primary shared construct here is the iterative nature of both processes.

The history of the ID field stems from attempts to effectively employ learning technologies such as slide shows and films in rapid training needs for wartime military learners. Thus, the field of ID was really established as a way to advance the use of hard technologies in the form of learning media. User-design, however, extends from a need to move design toward even beyond a service mentality to a point of empowerment—in which users become empowered to create their own learning. The history of user-design has been discussed more completely in chapter 2 and draws heavily from Scandinavian traditions. The impact of this difference between the two fields will probably only be known over time, but the alignment of these two fields with two very different value bases will likely lead to a significantly different path in their futures. I certainly hope, however, that despite this difference and others identified here, user-designers continue to exist at least tangentially within the larger Educational Technology field. Our presence within the field offers a strong source of energy, innovation, and attention to implementation and change, which has been, and continues to be, largely ignored in the field.

ID has been primarily applied in certain contexts with great success. For example, ID has been well-established within the military industrial complex and as part of the larger PT and training movement; ID has enjoyed strong success in corporations as well. ID has had slightly less success in public schools, where the impact has been debated (Carr & Reigeluth, 2001). Higher education has recently become particularly interested in ID and is increasingly employing instructional designers as they learn that this approach can help when moving courses from residential (face-to-face) instruction to online learning environments. Thus, the use of ID is on the rise in higher education. Few applications of ID have been effective within non-profits, though museums are perhaps the main exception to this generality. User-design is still very much in its infancy and has been primarily used in large, systemic design projects in public school systems. There are several cases being reported of user-design within schools that are facing dramatic changes, but fewer universities and only one or two cases of corporations implementing a user-design approach have been reported to date.

ID is embedded in the field of Instructional Systems/Educational Technology[3] (IS/ET) which is still an adolescent field in terms of its identity. As such, the field struggles with issues of identity and effectiveness. There have been many studies which have examined the identity of IS/ET as a field (IDT Futures Group, 2002; Jones, 1999; Reiber, 1998; Seels & Richey, 1994; Heinich, 1984; Silber, 1970) and task forces have been set up to help determine the field's core identities more definitively. Leaders in the field have tried to help IS/ET make sense of the field, who is "in" and who is "out" (Merrill, 1996), what counts as appropriate scholarship in the field, and what standards might be used to assess practitioners and novices in the field. A number of studies have looked at citations or publication outlets as ways to further understand our identity (e.g., Klein, 1997; Ross & Morrison, 1993; Ross, 1999; and Sachs, 1984).

User-design is much more in its infancy as a field and therefore is more focused on the very early stages of field definition. So, for user-design, we are most interested in what it is as a construct and process, where it comes from, what sort of initial attempts should be made in application of the user-design approach and sources of new ideas. The

---

[3]Many terms have been used to describe the field with a variety of qualifications regarding specific semantic differences between terms. Here we use *instructional systems–educational technology* as the most common term for the whole field, whereas *ID* is used to denote the instructional systems design process.

user-design field is less concerned with membership, standards, and boundaries and more concerned with inputs, new ideas, and initial applications.

The process of ID is organized around clearly discernable discrete steps. The traditional ADDIE (Analysis, Design, Development, Implementation, and Evaluation) model represents the most frequently cited and often utilized process that ID professionals employ in their own work. The process of user-design, on the other hand, is much more fluid and generally organized in loose phases. Understanding the details of this process difference is critical for you as a future ID user-designer as a first step in integrating the two. Therefore, I now turn full attention to the comparison of ID and user-design as processes.

ID & UD PROCESS COMPARISON

The processes involved in both ID and UD do follow basic stages, though to indicate any kind of linear progression might be a bit ambitious as far as the user-design process which I see as organized through several basic, loosely structured stages. Both processes are highly iterative in nature, but the ID process tends to have a more linear progression with opportunities for revision, rework, and revisiting particular steps and their outcomes. The UD process is a continually unfolding set of stages which are expected to be highly iterative. Table 7 compares the steps/stages in the two processes as they are ideally lived out. It is certainly the case that critics have suggested the ID process does not proceed so neatly as is expressed in almost all of its models. It is generally accepted that the ID process is far more messy

TABLE 7
Comparison of ID and UD Processes

ID Process Basic Steps	*UD Process Basic Stages*
Analysis	Readiness
Goal setting	Team selection
Selection of strategies	Process design and tool selection
Development of materials	Capacity building
Test development and assessment	Process engagement
Diffuse or adopt new instructional program	Trials of innovations
Assess student learning	Iterative assessment of process and product innovations
Evaluate instructional effectiveness	Evaluate UD systemic impacts

in the real world than it is in the idealized models of any theorist however, the existence of models gives us early tools for training novice instructional designers as well as a language to use to describe the process. Here I have tried to use a slightly modified generic model of the ID process which I hope encompasses most of the critical elements.

In both ID and UD there are some preparatory tasks which precede any serious design work. In the case of the ID process, full needs analyses, problem analyses, task analyses, etc. are usually the first steps. This helps to ensure that the problem is really a learning or training problem and that the need is best answered with an ID approach. In a very similar way, the user-design process begins with readiness assessments. While this includes an examination of the needs of the organization, the focus is also on the ability of the organization to be open to the power shifts and serious design work necessary in the UD approach.

In addition to readiness asessments, I consider the team selection process to be a part of the preparation for UD activities. This process certainly does take some time and is part of the reason that timelines are more protracted in this approach. Team selection in UD requires careful examination of size of team(s), number of team(s) if multiple work teams are desirable, and diversity within and across teams. Membership commitment and capacity/readiness of the individuals (as opposed to the organization that was determined in the first stage) to be involved is part of the team selection process as is leadership involvement/membership issues. The ID process begins in earnest by setting goals based on the needs of the organization discovered in the first step. These are typically learning or instructional goals at fairly broad levels and can run from very broad performance improvement goals to specific behavioral objectives related to the training itself. These two phases are really not equivalent or closely linked between the two processes.

The third step in the ID process is to align instructional strategies with the needs and goals for the training. Strategy selection is a broad category of many substeps and considerations such as motivational techniques, media selection, scope and sequence decisions, and instructional strategy choices. This is one of the more comprehensive steps and would be considered by most the "heart" of the design activity within the ID process. By the same token, the UD process engages in the creation of their own processes. As mentioned earlier, the UD approach is far more process oriented than the ID approach. So it is not a surprise, then, to see that the primary design activity in the early phases of the UD process is focused on deciding how design will proceed. Here the users determine what they want to adopt in terms of how they will engage in design, what values will guide them, and what tools (see chap. 3) they will use. Here again, because the user-design group might be working on process design while the ID team is working on product or ID, there is a sense in which the user-design process is more time consuming or lagging behind. The instructional design process will have some early designs that can be shared with leadership, while the UD process will have processes that can be shared, which may be far less impressive to leadership.

The fourth step/stage in these two approaches really illustrates the divergent paths that the two processes are taking at this point. About midway through the processes of UD and ID, we see very little overlap between the two. ID is concerning itself with the development of materials. During this step, graphic designers and producers are making powerpoint slides, videos are being produced, texts are being created. This is the nitty gritty product development stage—the "just do it" step. The user-design process, at this point, is focusing on the capacity building that is essential to user-design. Design is not an easy task, but it can be learned and participants in the user-design process are building a wide variety of design capabilities such as value setting, communication, conflict maximization, leadership, idea generation, and familiarity with discrete and continuous design events such as those outlined in Jenlink, Reigeluth, Carr & Nelson (1998) is all part of this fourth stage.

The fifth step in our generic ID process is the disciplined development of tests and assessment tools. Again, from a systems perspective it is essential that all the materials/decisions made in the ID process are in alignment with one another. Thus, the tests must measure the objectives that were set back in step two and reflect that content learned from the strategies in step three and the materials utilized in instruction itself from step four. Item analyses can help to ensure this alignment. Alternative assessment models such as portfolios, written products, on-the-job performance, and so forth would also be considered as possible assessment techniques in this stage. The UD process is now ready to really dig in to the design process itself. Meetings will take place and extensive discussions about various potential solutions will start percolating from among the users. Whatever particular tool was selected in the third step will be used now to collect data and analyze its usefulness to the team and the broader user group.

The ID process, in this sixth step is now ready to diffuse or implement the new program. Here is where the training actually takes place, new ideas are shared with the users for perhaps the first time. Users are now informed of the expected new behaviors and they are able to accept or reject the innovations/learnings either overtly or covertly. This is where most ID process fall down and fail to reach their intended goals. The user-design process is now prepared to begin trials of the innovations that they've been discussing. Primarily because of the user-design-based approach, the team will feel far more comfortable moving into rapid prototypes or trials of new ideas more seamlessly. Ideas are accepted or rejected based on how they really work on the frontlines and the feedback from within the user group. Here again is an opportunity for conflict among team members—conflict that should be maximized rather than avoided.

The seventh step for the basic ID process is to assess student learning. This is primarily accomplished through implementing the tests that were designed in the fifth step, but may also include on-the-job assessments and any alternative assessment techniques specified in the fifth step. Similarly, the UD process is engaged in the iterative assessment of both their processes and the resulting products or designs that they have been trying out during step six. There is really no beginning or end to this part of the process within the user-design approach and it is far less discrete than the corresponding assessments happening in the ID process.

The final step in the ID process is to evaluate the entire instructional program. This evaluation may serve formative goals of improvement or summative goals of decision making. A number of different measures may be employed to evaluate the instructional system including data collection and analysis and student feedback on the program. Too often this step is made up primarily of "smile sheets" in which the learners are invited to give their opinions of their instructors or the program materials, readings, activities and so on. More complete program evaluations include a variety of data collected and analyzed for performance indicators that align with the original organizational needs and goals set back in steps one and two. Thus the whole system feeds back upon itself and is both iterative and integrated. The final user-design step is very similar to the ID evaluation step in that the users examine the impact of their solutions on the organization. However, because the user-design initiative is embedded within the systemic change context, the evaluation examines a wide variety of data sources and takes a very broad look at the impact of user-design. Of particular attention in this step will be the adoption rates and the implementation of the new ideas resulting from user-design efforts as well as their impacts on the organization as a whole including the sub and supra systems and both anticipated and unanticipated ripple effects.

These two processes have some similarities and many differences in their specific process steps or stages. While the underlying values and approaches are vastly different, it is still very helpful to consider how the two compare since it can help to illuminate ways to integrate ID and UD together, which will be the topic of the final portions of this book. The comparison is rather like a path that begins together, forks off in different directions for awhile, and then joins again (see Figure 1). Some of the early stages are similar to one another, but during the middle part of the process, there are marked differences. Nearer to the completion of the process, the two come back together in their assessment and evaluation tasks. Ultimately the two are very different approaches and their differences are not necessarily highlighted by considering their corresponding steps or stages. Rather, it is through building the understandings that have been the focus of this book (understandings of empowerment, conflict, leadership, and tools) that will give you the most insights into the differences between these two approaches to the creation of human learning systems.

DAYS IN THE LIFE...

Perhaps one of the best ways to illustrate the realities of how these two approaches differ from one another and how they are, in fact, similar may be by simply discussing the daily activities that each practitioner may engage in when practicing these two diverse approaches.

A Day in the Life... ID

Franklin is an instructional designer working for a project manager (who has worked his way up through the organization) in a large retail organization with 28% of their market share. Stuff-It is a corporation with a large workforce and a strong active training department. The training department employs more than 100 people in various tasks associated with the production of training from soft skills such as team building, diverse workforce, and leadership, to more technical training for new employees and existing employees learning to use new technologies.

The organization follows a very traditional ID model for their primary approach to the creation of training and instructional materials and Franklin is involved in several projects at the same time. On Wednesday morning, Franklin comes to work and scans his email and snail mail. He finds several e-mails from Camille, the graphic artist, asking him for feedback on shelf stocking skills training handouts. He reviews them and offers feedback by email. He also finds, in his snail mail several memos from his boss regarding new training projects on

LINKING USER-DESIGN 117

the horizon. The training group does an excellent job of keeping track of their tasks at least six months out.

One of the memos mentions a soft skills training design/development project which is slated for implementation in eight months time. The focus of the training is leadership for inspiration and is targeted at management trainees. Franklin makes some notes on the memo to remind himself of some ideas he has around this particular new program and files it in his "ideas" file.

Before he knows it, time has crept up on him and his first morning meeting at 9 a.m. is ready to start. This meeting is a status report on the progress being made on the new cash register technology training, one of the largest new technology training programs in the entire organization. He has roughed out the needs statement and shares it with the team at the morning meeting. There is general consensus on the need for this training and, since the focus is pretty narrowly on the specific skills needed and not the more systemic issues that may arise as a result of the new cash register systems, everyone pretty well accepts Franklin's work as expert and they finalize it with just a little wordsmithing.

After the cash register training design team meeting, Franklin spends several hours with a subject matter expert on store-level accounting procedures writing out a task analysis, task taxonomy, and then works on his own on translating these analyses into behavioral objectives.

At lunch, Franklin spends some time with his colleague Danielle who works in graphics. They discuss the drafts that Camille sent Franklin for the shelf stocking skills training, and they generate a few new ideas. After lunch, Franklin realizes that these ideas aren't really "approved" and haven't been included in the behavioral objectives. He will either have to go through a fairly long work process to get the new ideas that implied new content approved or ditch it. After a brief time wrestling with his own numbers to figure out the potential cost–benefit, he decides to ditch the cafeteria borne ideas and emails the graphic artist.

The entire afternoon is spent in a workshop to discuss a problem that has come up with the existing employee benefits training. The training was intended for current (not new) employees to explain and describe a number of changes in their benefits. The purpose of the training was to help current employees learn more about how to report sicknesses, vacation days, and overtime. The forms have changed for health benefits reporting, and employees needed to learn how to complete them and who to submit them to. In addition, travel procedures have changed radically over the past three years time and employees are being asked to substantially change the procedures that they had been using for finding cheap airfares. The trouble is that now that the program is in the implementation phase, the training department has discovered that the trainees are spending almost all of the training time period discussing the changes. They have been complaining about the problems that this will cause them individually and the implications for the changes system wide. They are rarely getting to the specific steps for filing paperwork or changes to forms, and two of the training sessions have been so contentious that there has been talk among the workers of unionization.

No one in the training department had anticipated anything like this. What had been seen as a pretty simple and straight-forward skills training program aimed at helping current employees has become a groundswell of anticorporate sentiment. The workshop starts by having the CEO of the entire company come to the training department charging them with the task of communicating this information without allowing for this sort of negative communication across workers. The session is meant to be a brainstorming session where some better solution will result from the training department. After several hours it is decided that an alternative to training might be a best solution given the problems that are resulting at the instructional moment. The training department recommends that "Morning Briefings" be issued once a week outlining these and similar changes in the future along with flowcharts and specific job aids to help current employees use the new procedures. Lots of good ideas follow the initial disappointment that the training itself was resulting in such turmoil, and the morning briefings are assigned for mockup to the graphics department with content to be pulled by the instructional designers from the current unsuccessful training program.

Franklin spends the last hour of his day on his e-mail catching up on the goings-on of the afternoon. He receives an evaluation report on the well-received clothing buyers' training program pilot and scans it noting that the costs were higher than anticipated, but the goals seem to have been largely met, and the learners were happy. Feeling a bit better about life after this bit of good news, he decides it is time to pack it in for the day and heads home.

A Day in the Life . . . User-Design ID

Joy hums as she drives into the parking lot at My Things! where she works as an ID–UD facilitator. Her official title on her business card reads, "Training Facilitator," but she feels like she is really a human performance improvement facilitator. My Things! is very similar to

LINKING USER-DESIGN 119

Stuff It in terms of its market and business model. My Things! does have a 29% market share, so the two companies are in pretty tight competition, but they are both the dominant dogs in the arena.

My Things! has decided to take some risks on some new business ideas. They have been early technology adopters, at times failing but learning much from their failures. They have a very family/team oriented employee model, calling them all *partners* rather than *workers*. Some find the orientation a little cheesy, but Joy really likes working for My Things! because she knows that what she does matters to the other partners and to the organization as a whole.

Joy gets to her work station and checks her emails and snail mail. So much junk mail and spam, Joy wades through to find the "real" stuff. She finds a note from Sandra, head of a user-design team currently working through some new ideas for greeter training. Sandra is looking for some additional information from leadership in terms of costs and benefits of greeters to help the team make sense of some of their latest notions. Joy, as the facilitator, is happy to oblige by making the connection between Sandra and Alex, a member of the leadership team who is most likely to have that accounting-type information. She calls Alex first to make sure that he's a good source for the information and then that he's able to get and share it. Having confirmed this, she lets him know that Sandra will be in touch shortly. The whole thing could have been done with a few e mails, but Joy prefers high touch in her facilitation whenever possible. Some teams need more hands-on and others need less. My Things! has been pretty good about matching up facilitation styles with team needs through the last three years of user-design practice.

Joy hurries off to her 9 a.m. meeting with the capacity building team. They are discussing the needs of three new user-design teams with their facilitators today. Fortunately, none of her teams are in the initial capacity building stages, so she's able to brainstorm with others about their work with returns agents, truck drivers, and staff assistants at central office. These three teams are working on various innovative processes and products to improve their own lives and performance for the organization.

After the meeting, she logs onto the video *isight* camera. She loves this way of connecting with her teams; though she'd prefer to be in person at each meeting, it is financially impossible to do so. The new Tiger operating system for her Macintosh laptop allows her to connect with her team and two or three others as needed. She has a meeting first with the user-design team for the new cash register training. Everyone in the industry is making this transition, she knows. And at times she wonders if this was the best use of user-design after all, she's not completely convinced that the need for new ideas here is very high, but it's become a way of life for the users who have been working in a user-design team for more than 1 year together now, so putting the new cash register training through them really only makes sense. Conversely, *NOT* putting it through them is likely to raise red flags.

The team is currently in the tool selection stage, and they consider several options before settling on a cooperative design approach with a special interest in the possibility of cooperative prototyping with the cash register software company itself. Rather than simply accepting the expertise of the vendor, the users decide that they want a hand in what their interface is going to look like, and Joy fully supports this notion.

Her next meeting is in person, just across the courtyard working with a user-design team in accounting. They have been struggling for some time with overtime pay accounting procedures and have decided on ethnographic data collection to learn more about how everyone in accounting is dealing with the overtime issue. They have collected their data and are in the midst of making sense of it. Joy's primary purpose at this point in the process is to ensure that nothing gets in the way of their forward motion and that they have as much information as possible. Joy's not all that happy with their progress, this team seems to be a little slower than would be optimal in terms of moving ahead. If they need additional documents or already extant data, it is Joy's job to be sure they gain access to it in a timely fashion. There's not a lot that Joy can do to move things along, though she'd like to just take charge and prescribe some next steps for them. Sometimes she has to simply bite her tongue and let the process go where it will.

She has one more isight meeting prior to lunch in which a user-design team in marketing is working on ways to improve their reporting and communication processes within the marketing departments internationally. This team chose a somewhat more traditional path with a basic needs assessment set of tools to help them pinpoint their needs and performance gaps. However, the actual assessment was done by members of the team, each of whom took responsibility for different parts of the report. The first report from this assessment is coming in today, and Joy is really proud to hear about their progress. It seems pretty clear that most of the information that they've gotten jibes with their own experiences, and she offers them a few more readings on needs assessments because they seem to have missed one of the major steps, task analysis, and Joy wants them to consider whether a structured approach to task analysis would be helpful in this instance.

Lunchtime is shared with her colleagues from across several departments, they are a group of women she has a book club with. Fortunately today was not a book club day, though, as Joy really doesn't

LINKING USER-DESIGN 121

know how she would have had time to read any part of the new book *Linked* before lunch!

After lunch Joy meets with several corporate leaders to report on the progress of several of her user-design teams. The remainder of the afternoon is spent working with a new team that wants to form in order to design a response to the changes happening in the benefits area. The first steps in this regard are determining the extent to which UD makes sense for this task and so this team, which has formed from the grassroots, starts by having several meetings (this is their third) to discuss what it is about the changing benefits environment that they feel deserves a response and the extent to which they meet the criteria set out for user-design teams. Their focus today is on the systemic implications of the benefits changes as a justification for user-design.

Joy decides to spend a little more time working on some of the needs/task analysis readings to make sure that she's given the user-design team in marketing sufficient and appropriate information for their work then it's quittin' time.

SUMMARY AND THOUGHT QUESTIONS

These two stories of daily life illustrate clearly, I believe, the differences and sameness in the work that a user-designer does as compared to a traditional instructional designer. There is a good bit more detailed work and in some cases more accountable results from the traditional instructional design professional. However, the traditional training department is having its moments with adoption and implementation, and the story there is probably not one that is at all that unfamiliar to those engaged in traditional instructional design in training departments all across the country.

The user-design tale is one of ups and downs, and just like the traditional ID story, it details its own share of problems. Joy isn't convinced in all cases that user-design is the solution, and some of her teams seem to be "slow as molasses" in getting anything done. She has to account for the UD team progress and, in most cases, it is not at all under her control. Facilitation is not the same as control, and so there is an inherent conflict here that is problematic for Joy. But, she also recognizes that the true improvement in human performance, the empowerment that her team members enjoy, and the advances that the organization are making system wide will help to ensure a healthy organization and jobs for these workers for many years to come.

The user-design story probably sounds a bit more unlikely or ideal. Perhaps the idea is that the organization has accepted and sustained the user-design team and is open to much innovation. So, this is unrealistic just now because very few organizations have even tried user-design much less integrated it as a model for how they do business, but imagining what the idea looks like helps to illustrate how the processes and what people do within them really differ and how they are similar. In reality, the user-design approach has not been used to such an extent in any organization. Schools are coming closer with more and more openness to innovation and innovative processes. But, stories of daily life in schools would probably look very different than these narratives.

Exploring the differences between ID and user-design, the way the lived experience differs or is the same, the way the process stages or steps differ or are the same, and integrating these ideas into more user-design-oriented instructional design models is probably the next logical step for readers of this book. It is time now for you to go out into the world with the knowledge of the limitations of traditional instructional design approaches and imbue those models with innovative ways to approach training, learning, instruction, and design that empower the workers, the learners, the teachers, those who are closest to the problem.

> What do you think you would like most/least about Joy's job?
> How do you see the linkage between ID and UD? Draw a picture if you can.
> What are the most critical shared elements?
> How do you see the comparison of the ID and UD processes/stages? Draw a picture if you can.

CONCLUSION

Users, who are the users? It is not at all hard to find users it's just hard to see them. I have three small children—the children to whom this book is dedicated. They are users. They use many things in my world: They use the house we live in, they use the brightly colored eating utensils we bought them with little animals on the handles, they use various physical artifacts, and then many more social systems set up in our home. They are users of routine; we have a routine schedule for them, and they use it. They are users of parents, sounds a bit odd, doesn't it, but they are. They use us to develop their abilities to see the world to get along with their siblings and others in the world—in short, to become properly socialized. They use us to learn languages, letters, numbers, and sciences, and all about their world by asking us questions and conversing with us.

I have discovered that users are like systems. Once you open your eyes and you open yourself to seeing them, they are everywhere. It's almost debilitating to realize that you cannot ignore the users, nor can you fix things for them; rather they have to become independent users who can create and design for themselves, their systems, their futures, their lives. All these things are constructed; they do not just happen to us; they are socially negotiated among our whole human culture, and they are created by us for many purposes.

Designing instruction and human learning environments can be approached from a linear and systematic mindset; it can be imposed on the learners with set objectives and with clearly aligned strategies, assessments, and activities. However, I fear that our past failures when this approach is formulaically applied tell us much about the merits of considering another approach, or softening that approach with some sense of UD—moving design toward empowered service. As an instructional designer you hold in your hands great power to move the field forward into an age in which we respect and empower each user to create the sort of future they need and want, to create learning systems that truly work for them and that are sustainable through relatively effortless adoption and implementation processes. It's not an easy path to say the least. Indeed, taking on the charge of user-design is incredibly difficult and overwhelming. Overcoming disagreements, conflict, unclear systemic purposes, dynamism, and constant change can be exhausting at times. But the fight is worth the payoff. As a designer who is dedicated to the moral purpose of empowering users, to serving them and helping them to create the kind of learning that they want, you will find yourself engaged in some of the most exhilarating and rewarding design work of your life.

References

Abels, Eileen. G., White, Marilyn Domas, Hahn Karla (December 1997). *The development and implementation of user-based design process in Web site design* (Disclosure, Inc. Rep. No. CLIS-TR-97-04). Maryland Univ, College Park.

Adler, P. A., & Adler, P. (1994). Observational techniques. In N. Denzin & Y. Lincoln (Eds.), Handbook of qualitative research (pp. 377–392). Thousand Oaks, CA: Sage.

AML. (1977). *Law of workers' protection and working environment.* Oslo, Norway: Anti Money Laundering

Apple, M. W. (1986). *Teachers and texts: A political economy of class and gender relations in education.* New York: Routledge.

Apple, M. W. (1988). Teaching and technology: The hidden effects of computers on teachers and students. In L. E. Beyers & M. W. Apple (Eds.), *The curriculum: Problems, politics, and possibilities* (pp. 289–311). Albany: State University of New York Press.

Apple, M. W. (1990). *Ideology and criticism.* New York: Routledge.

Argyris, C., Putnam, R., & Smith, M. C. (1985). *Action science: Concepts, methods, and skills for research and intervention.* San Francisco: Jossey-Bass.

Atkinson, P., & Hammersley, M. (1994). Ethnography and participant observation. In N. Denzin & Y. Lincoln (Eds.), *Handbook of qualitative research* (pp. xxx–xxx). Thousand Oaks, CA: Sage.

Banathy, B. H. (1973). *Developing a systems view of education: The systems-model approach.* Belmont, CA: Siegler/Feraon.

Banathy, B. H. (1991). *Educational systems design: A journey to create the future.* Englewood Cliffs, NJ: Educational Technology Publications.

Banathy, B. H. (1992). Comprehensive systems design in education: Building a design culture in education. *Educational Technology, 22*(3), 33–35.

Banathy, B. H. (1996). *Designing social systems in a changing world.* New York: Plenum.

Banathy, B. H. (2003). *The evolution of systems inquiry. The Primer Projects of International Society for the Systems Sciences (ISSS).* Retrieved June 2, 2005, from http:/www.isss.org/primer/data/003evsys.htm

Bansler, J. (1989). Systems development research in Scandinavia: Three theoretical schools. *Scandinavian Journal of Information Systems, 1,* 3–20.

Barab Sasha A. and Squire, Kurt (2004) Design based research: putting a stake through the ground. Journal of Learning Sciences, 13, pp. 1–14

Barabasi, A. L. (2003). *Linked: How everything is connected to everything else and what it means.* Penguin Ltd. New York: Plume.

Barrows, H. S. (1986). A taxonomy of problem-based learning methods. *Medical Education, 20,* 481–486.

Bateson, G. (1979). *Mind and nature, a necessary unity.* New York: Dutton.
Bauch, P. A., & Goldring, E. B. (1998). Parent–teacher participation in the context of school governance. *Peabody Journal of Education, 73,* 73–98.
Bell, A., & Forbes, R. (1997). Performance management. In R. Kaufman, S. Thiagarajan, & P. MacGillis (Eds.), *The guidebook for performance improvement: Working with individuals and organizations* (pp. 371–392). San Francisco, CA: Jossey-Bass.
Berube, M. (1970). Community control: Key to educational achievement. *Social Policy, 1,* 42.
Bhola, H. S. (1977). *Diffusion of educational innovation.* Morristown, NJ: General Learning Press.
Bjerknes, G., & Bratteteig, T. (1995). User participation and democracy: A discussion of Scandinavian research on system development. *Scandinavian Journal of Information Systems, 7,* 73–98.
Blomberg, J., Giacomi, J., Mosher, A., & Swenton-Wall, P. (1993). Ethnographic field methods and their relation to design. In D. Schuler & A. Namioka (Eds.), *Participatory design: Principles and practices* (pp. 123–156). Hillsdale, NJ: Lawrence Erlbaum Associates, Inc.
Bodker, S. (1996) Applying activity theory to video analysis: how to make sense of videodata in human-computer interaction. In Nardi, B.A. (Ed.) Context and Consciousness: Activity theory and human computer interaction, MIT press, Cambridge, M.A. pp. 147–174.
Bødker, S., Gronbaek, K., & Kyng, M. (1993). Cooperative design: Techniques and experiences from the Scandanavian scene. In D. Schuler & A. Namioka (Eds.), *Participatory design: Principles and practices* (pp. 157–175). Hillsdale, NJ: Lawrence Erlbaum Associates, Inc.
Bohm, D. (1980). *Wholeness and the implicat order.* London: Ark. London: Routledge and Kogan Page.
Bohm, D. (1990). *On dialogue.* Ojai, CA: David Bohm Seminars.
Boulding, K. E. (1985). *The world as a total system.* Beverly Hills: Sage.
Bromley, H. (1992). Culture, power, and educational computing. In C. Bigum & B. Green (Eds.), *Understanding the new information technologies in education: A resource for teachers* (pp. xxx–xxx). Geelong, Australia: Deakin University Press.
Brown, J. S., & Duguid, P. (1994). Practice at the periphery: A reply to Steven Tripp. *Educational Technology, 34*(8), 9–11.
Burns, J. (1978). *Leadership.* New York: Harper & Row.
Capra, F. (1982). *The turning point: Science, society and the rising culture.* New York: Bantam.
Carr-Chellman, A.A., & Hoadley, C.M. (2004) Introduction to special issue: Learning Sciences and instructional systems: Beginning the dialogue. *Educational Technology, 44*(3), 5–6.
Carmen, R. (1990). *Communication, education and empowerment* (Report No: ISBN-0-902252-20-8). England: Manchester University, Center for Adult and Higher Education. (ERIC Document Reproduction Service No. ED362649).
Carr, A. A. (1994). Community participation in systemic educational change. *Educational Technology, 34*(1), 43–50.
Carr, A. A. (1995). Stakeholder participation in systemic change: Cornerstones to continuous school improvement. In P. Jenlink (Ed.), *Systemic change: Touchstones for the future school* (pp. 71–85). Palatine, IL: Skylight.
Carr, A. A. (1996). Distinguishing systemic and systematic! *Tech Trends, 41,* 16–20.
Carr, A. A. (1997). User-design in the creation of human learning systems. *Educational Technology Research and Development, 45*(3), 5–22.

REFERENCES

Carr, A. A., & Reigeluth, C. M. (2001). Whistling in the dark? ID in the schools. In R. A. Reiser & J. Dempsy (Eds.), *Trends and issues in instructional design and technology* (pp. xxx–xxx). Englewood Cliffs, NJ: Merrill/Prentice Hall.
Carr, I. C. (1990). The politics of literacy in Latin America. *Convergence, 23*(2), 50–68.
Carr-Chellman A.A. (December, 1998) Systemic change: Critically reviewing the literature. Educational Research and Evaluation, 4 (4) pp. 369–394
Carr-Chellman, A. A., & Hoadley, C. M. (2004). Introduction to special issue: learning sciences and instructional systems: Beginning the dialogue. *Educational Technology,* 44(3), 5–6.
Carr-Chellman, A. A., Cuyar, C., & Breman, J. (1998). User-design: A case application in health care training. *Educational Technology Research and Development, 46*(4), 97–114.
Carr-Chellman, A.A., & Savoy, M.S. (2003). Using the user-design research for building school communities. The School Community Journal, 13(2), 99–118.
Carroll, J. M., & Rosson, M. B. (1992). Getting around the task–artifact cycle: How to make claims and design by scenario. *ACM Transactions on Information Systems, 10,* 181–212.
Carspecken, P. F. (1996). *Critical ethnographyin educational research: A theoretical and practical guide.* New York: Routledge.
Checkland, P. (1999). *Systems thinking, systems practice.* Chichester, England: Wiley.
Chermack, T.J., Lynham, S.A. Ruona, W.E.A. (2001) A review of scenario planning literature. Future Research Quarterly 17(2) pp. 7–31.
Chomsky, N. (1998) Propaganda and control of the public mind. In R. W. McChesney, E. M. Wood, & J. B. Foster (Eds.), *Capitalism and the information age: The political economy of the global communication revolution* (pp. 179–190). Montly Review Press (June 1997)
Churchman, C. W. (1968). *The systems approach.* New York: Delacorte.
Cochran, M., & Dean, C. (1991). Home–school relations and the empowerment process. *The Elementary School Journal, 91,* 261–269.
Cognition and Technology Group at Vanderbilt. (1993). Anchored instruction and situated cognition revisited. *Educational Technology, 33*(3), 52–70.
Collins A., Joseph, D. Bielaczyc (2004) Design research: theoretical and methodological issues. Journal of the learning sciences.
Comer, J. P., & Haynes, N. M. (1991). Parent involvement in schools: An ecological approach. *The Elementary School Journal, 91,* 271–277.
Comstock, D. E., & Fox, R. (1993). Participatory research as critical theory: The North Bonnevile, USA, experience. In P. Park, M. Brydon-Miller, B. Hall, & T. Jackson, (Eds.), *Voices of change: Participatory research in the United States and Canada* (pp. 103–125). Westport, CT: Bergin & Garvey.
Cooper, B. S. (1992). A tale of two cities: Radical school reform in Chicago and London. In J. J. Lane & E. G. Epps (Eds.), *Restructuring the schools: Problems and prospects* (pp. 109–121). Berkeley, CA: McCutchan.
Coupland, D. (1995). *Microserfs.* New York: Regan.
Damarin, S. K. (2001). Technology and multicultural education: The question of convergence. *Theory Into Practice, 37,* 11–19.
Daniels, D. H., Kalkman, D. L., & McCombs, B. L. (2001). Young children's perspectives on learning and teacher practices in different classroom contexts: Implications for motivation. *Early Education and Development, 12,* 253–273.
Daresh, J. C. (1992). Impressions of school-based management: The Cincinnati story. In J. J. Lane & E. G. Epps (Eds.), *Restructuring the schools: Problems and prospects* (pp. 109–121). Berkeley: McCutchan.
Davidson, M. (1983). *Uncommon sense, the life and thoughts of Ludwig von Bertalanffy.* Boston: Houghton Mifflin.

Davies, D. (1981). Citizen participation in decision making in the schools. In D. Davies (Ed.), *Communities and their schools* (pp. 83–119). New York: McGraw-Hill.

Davydov, V. V. (1995). The influence of L. S. Vygotsky on education theory, research and practice. *Educational Researcher, 24*(3), 12–21.

DeCandido, G. A. (1997). *After the user survey, what then? Issues and innovations in transforming libraries*. Washington, DC: Systems and Procedures Exchange Center.

Delgado-Gaitan, C. (1991). Involving parents in the schools: A process of empowerment. *American Journal of Education, 100*, 20–46.

Demming, W. E. (1982). *Out of the crisis*. Cambridge, MA: IT Center for Advanced Engineering Study.

Design-Based Research Collective. (2003). Design-based research: An emerging paradigm for educational inquiry. *Educational Researcher, 32*(1), 5–8.

Dick, W. Carey, L. & Carey, J. (2005) The systematic design of instruction (6th ed.). New York: Harper Collins

Dick, W., & Carey, L. (1996). *The systematic design of instruction* (4th ed.). New York: HarperCollins.

Dick, W., & Johnson, F. C. (1993). Quality systems in performance improvement [Special issue]. *Performance Improvement Quarterly, 6*(3).

Dillon, A. (1995). Artifacts as theories: Convergence through user-centered design. *Proceedings of the ASIS Annual Meeting, 32*, 208–210.

Dreikorn, M. J. (2004). *The synergy of one: Creating high-performing sustainable organizations through integrated performance leadership*. Milwaukee, WI: Quality Press.

Duffy, T. M., Lowyck, J., & Jonassen, D. H. (1993). *Designing environments for constructive learning*. Berlin: Springer-Verlag.

Elden, R. (1979) Three generations of worker democracy research in Norway. In C.L. Cooper and E. Mumfords (Eds.) The quality of work handbook for action. Thousand Oaks, C.A: Corwin Press

Epstein, J. L. (1997). *School, family, and community partnerships: Your handbook for action*. Thousand Oaks, CA: Corwin.

Ertmer, P. (2000, Oct.). *Responsive insructional design: Scaffolding the adoption and change process*. Paper presented at the 2000 annual convention of the Association of Educational Communications and Technology, Denver, CO.

Evans, R. (1996). *The human side of change: Reform, resistance, and the real-life problems of innovation*. San Francisco: Jossey-Bass.

Fantini, M., Gittell, M., & Magat, R. (1970). *Community control and the urban school*. New York: Praeger.

Fidel, R. (1994). User-centered indexing. *Journal of the American Society for Information Science, 45*, 572–576.

Flood, R. L. (1990). *Liberating systems theory*. New York: Plenum.

Flood, R. L., & Romm, N. R. A. (Eds.). (1996). *Critical systems thinking: Current research and practice*. New York: Plenum Press.

Fontana, A., & Frey, J. H. (1994). Interviewing: The art of science. In N. Denzin & Y. Lincoln (Eds.), *Handbook of qualitative research* (pp. 361–376). Thousand Oaks, CA: Sage.

Ford, J. K., (1968/1999). Organizational development. In D. G. Langdon, K. S. Whiteside, & M. M. McKenna (Eds.), *Intervention resource guide: 50 performance improvement tools* (pp. 251–258). San Francisco, CA: Jossey-Bass.

Freire, P. (1970).0 *Pedagogy of the oppressed* (M. B. Ramos, Trans.). New York: Seabury Press. (Original work published 1968/1970)

Gagne, R. M., Briggs, L. J., & Wager, W. W. (1992). *Principles of instructional design*. Fort Worth, TX: Harcourt Brace.

Gardner, J. W. (1963). *Self-renewal*. New York: Harper & Row.

REFERENCES

Garrison, J. (1995). Deweyan pragmatism and the epistemology of contemporary social constructivism. *American Educational Research Journal, 32,* 716–740.

Gayeski, D. (1995). The changing role of human performance technology [Special issue]. *Performance Improvement Quarterly, 8*(2).

Gerhardt, H. P. (1986). *Brazil's popular education in the eighties: Essentials, fundamentals and realpolitik.* Verlag für Interkulturelle Kommunikation, Frankfurt (ERIC Document Reproduction Service No. ED291877)

Gilbert, T. F. (1978). *Human competence: Engineering worthy performance.* New York: McGraw-Hill.

Giroux, H. A. (1992). Educational leadership and the crisis of democratic government. *Educational Researcher, 21*(4), 4–11.

Grudin, J. (1993) Obstacles to participatory design in large product development organizations in participatory design: principles and practices. D. Schuler and A. Namioka (Eds.) Hillsdale N.J. Erlbaum, pp. 99–119

Habermas, J. (1984). *The theory of communicative action: Vol. 1, reason and the rationalization of society.* Boston: Beacon.

Habermas Jurgen (1987) The philosophical discourse of modernity: twelve lectures. Translated by Fredrick G. Lawrence Cambridge, MA, MIT Press.

Habermas, J. (1990). *Moral consciousness and communicative action.* Cambridge, MA: MIT Press.

Hall, A. D. (1962). *A methodology for systems engineering.* Princeton, NJ: Van Nostrand.

Hannum, W. (2005). Instructional systems development: A 30 year retrospective. *Educational Technology, July/August,* 5–20.

Harman, W. W. (1984). How I learned to love the future. *World Future Society Bulletin, 18*(6), 1–5.

Heidegger, M. (1977). The question concerning technology. In D. F. Krell (Ed.), *Martin Heidegger: Basic writings* (pp. 283–317). San Francisco: HarperCollins.

Heinich, R. (1984). The proper study of instructional technology. Educational Communications and Technology Journal, 33(1), 9–15.

Henderson, L. (1996). Instructional design of interactive multimedia: A cultural critique. *Educational Technology Research and Development, 44*(4), 85–104.

Hirschkop, K. (1998). Democracy and the new technologies. In R. W. Mc Chesney, E. M. Wood, & J. B. Foster (Eds.), *Capitalism and the information age: The political economy of the global communication revolution* (pp. 207–217). The Monthly Review (publisher)

Hitch, C. J. (1955). An appreciation of systems analysis. In S. L. Optner (Ed.), *Systems analysis* (pp. 466–481). Harmondsworth, England: Penguin.

Hlynka, D. (1996). Postmodernism. In A. R. J. Yeaman, D. Hlynka, J. H. Anderson, S. K. Damarin, R. Muffoletto (Series Eds.), & D. H. Jonassen (Vol. Ed.), *Handbook of research for educational communications and technology* (pp. 253–265). Mahwah, NJ: Lawrence Erlbaum Associates.

Hoadley, C. M. (2004). Methodological alignment in design-based research. *Educational Psychologist, 39,* 203–212.

Hoadley, C. M. (2005). Design-based research methods and theory building: A case study of research with SpeakeEasy. *Educational Technology, 45*(1), 42–46.

Holloway, L. D. (1972). The learner-centered approach to instruction. *American Vocational Journal, 47,* 32–34, 53–55.

Horkheimer, M., & Adorno, T. W. (2002). *Dialectic of enlightenment: Phiosophical fragments* (E. Jephcott, Trans.). Stanford, CA: Stanford University Press. (Original work published 1944/2002)

Horn, R. A., & Carr, A. A. (2000). Changing schools systemically: The use of moral conversation in professional development. *Systems Research and Behavioral Sciences, 17*(3), 23–38.

Horton, B. D. (1993). The Appalachian land ownership study: Research and citizen action in Appalachia. In P. Park, M. Brydon-Miller, B. Hall, & T. Jackson, (Eds.), *Voices of change: Participatory research in the United States and Canada* (pp. 85–102). Westport, CT: Bergin & Garvey.

Hurley, D. E. (1980). A method for gathering user input to achieve a successful design system. *Cause/Effect, 3*(3), 22–27.

Hutchins, C. L. (1996) *Systemic thinking: Solving complex problems.* Aurora, CO: Professional Development Systems.

IDT Futures Group (2002). Assessing the field of educational technology: What is the state of the field? Presentation to the Annual Convention of the Association of Educational Communication & Technology (AECT), October.

ISSS (International Society of Systems Sciences). (2003a). *Characteristics of a human activity system.* Retrieved June 2, 2005, from http://drupal.ieeecleveland.org/?q=book/print/38

ISSS (International Society of Systems Sciences). (2003b). *Conscious evolution of humanity: Using systems thinking to construct agoras of the global village.* Retrieved June 2, 2005, from http://www.democracy.nildram.co.uk/new_visions/systems_thinking/isss_intro.htm

Jackson, T. (1993). A way of working: Participatory research and the aboriginal movement in Canada. In P. Park, M. Brydon-Miller, B. Hall, & T. Jackson (Eds.), *Voices of change* (pp. 243–260). Westport, CT: Bergin & Garvey.

Jantsch, E. (1980). *The self-organizing universe: Scientific and human implications of the emerging paradigm of evolution.* New York: Pergamon.

Jenkins, G. M. (1972). The systems approach. In J. Beishon & G. Peters (Eds.), *Systems behavior* (p. 73). London: Open University Press.

Jenlink, P. M. (1995). *Systemic change: Touchstones for the future school.* Palatine, IL: Skylight.

Jenlink, P. M., Reigeluth, C. M., Carr, A. A., & Miller, L. M. (1998). Guidelines for facilitating systemic change in school districts. *Systems Research and Behavioral Science, 15,* 217–233.

Jonassen, D. H. (1991). Objecitivism versus constructivism: Do we need a new philosophical paradigm? *Educational Technology Research and Development, 39*(3), 5–14.

Jonassen, D. H. (1999). Designing constructivist learning environments. In C. M. Reigeluth (Ed.), *Instructional-design theories and models: A new paradigm of instructional theory* (Vol. 2, pp. 215–239). Mahwah, NJ: Lawrence Erlbaum Associates, Inc.

Jones, B.W. (1999). A differentiating definition of instructional technology and educational technology. Retrieved 24 May, 2005 from www.geocities.com/capecanaveral/campus/7941/trmpp.rh.html.

Jones, J. C. (1970). *Design methods: Seeds of human futures.* New York: Wiley.

Kane, E. (1997) Participatory rural appraisal for educational research: Helping to see the invisible. Irish Journal of Anthropology 2, pp. 69–85

Kapoor, I. (2002) The devil's in the theory: A critical assessment of Robert Chamber's work on participatory development. Third World Quarterly, 23, pp. 101–117

Kelley, R. (1992). *The power of followership.* New York: Doubleday.

Kelly, K. (1994). *Out of control: The new biology of machines, social systems, and the economic world.* Reading, MA: Addison-Wesley.

Klein, J.D. (1997) ETR & D-Development: An analysis of content and survey of future directions. Educational Technology Research and Development, 45(3), 57–62

Kovach, K.A. (1987) Practical Labor Relations. University Press of America (p. 154)

Land, S. M., & Hannafin, M. J., (1996). A conceptual framework for the development of theories-in-action with open-ended learning environments. *Educational Technology Research and Development, 44*(3), 37–53.

REFERENCES

Larsen, T. J., & McGuire, E. (1998). *Information systems innovation and diffusion: Issues and directions.* Hershey, PA: Idea Group.

Laurillard, D. (2002). *Rethinking university teaching: A conversational framework for the effective use of learning technologies* (2nd ed.). New York: Routledge.

LeCompte, M. D., & Preissle, J. (1993). *Ethnography and qualitative design in educational research.* New York: Academic.

Lineberry, C., & Carleton, J. R. (1992). Culture change. In H. D. Stolovitch & E. Keeps (Eds.), Handbook of human performance technology: A comprehensive guide for analyzing and solving performance problems in organizations (pp. xxx–xxx). San Francisco: Jossey-Bass.

Lippett, L. (1998). *Preferred futuring: Envision the future you want and unleash the energy to get there.* San Francisco: Barrett-Koehler.

MacDonald, B. (1971). Briefing decision makers: Evaluation unit of the humanities curriculum project. (Reprinted in *School evaluation: The politics and process,* pp. 174–187, by E. House, Ed., 1973, Berkeley, CA: McCutchan)

Maher, F. A., & Tetreault, M.K.T. (1994). *The feminist classroom.* New York: Basic Books.

McCandless, P., et al. (1985). *University of Illinois library. The invisible user: User needs assessment for library public services.* Washington, DC: Association of Research Libraries.

McChesney, R. W. (1998). The political economy of global communication. In R. W. McChesney, E. M. Wood, & J. B. Foster, (Eds.), *Capitalism and the information age: The political economy of the global communication revolution* (pp. 1–26). XXcity, state: publisherXX.

McCombs, B. L. (2001). What do we know about learners and learning? The learner-centered framework: Bringing the educational system into balance. *Educational Horizons, 79,* 182–193.

McCombs, B. L., & Quait, M. (2000). *Results of pilot study to evaluate the community for learning (cfl) program.* Philadelphia, PA: Mid-Atlantic Lab for Student Success.

McCombs, B. L., & Whistler, J. S. (1997). *The learner-centered classroom and school: Strategies for increasing student motivation and achievement.* San Francisco, CA: Jossey-Bass.

McLaren, P. (1994). Multiculturalism and the postmodern critique: Toward a pedagogy of resistance and transformation. In H. A. Giroux & P. McLaren (Eds.), *Between borders: Pedagogy and the politics of cultural studies* (pp. 192–224). Routledge.

Melo, A., & Benavente, A. (1978). *Experiments in popular education in Portugal* (ERIC Document Reproduction Service No. ED182086)

Merrifield, J. (1993). Putting scientists in their place: Participatory research in environmental and occupational health. In P. Park, M. Brydon-Miller, B. Hall, & T. Jackson (Eds.), *Voices of change* (pp. 65–84). Wesport, CT: Bergin & Garvey.

Merill, M.D., Drake, L. Lacy M.J. & Pratt, J. (1996) Reclaiming instructional design. Educational Technology, 36(5), 5–7

Midgley, C., & Wood, S. (1993). Beyond site-based management: Empowering teachers to reform schools. *Phi Delta Kappan, 75,* 245–252.

Miller, J. G. (1978). *Living systems.* New York: McGraw-Hill.

Ming-fen, L. (2000). *Fostering design culture through cultivating the user-designer's design thinking and systems thinking.* Denver, CO: Association for Educational Communications and Technology. (ERIC Document Reproduction Service No. ED455775)

Morris, R. C. T. (1994). Toward a user-centered information service. *Journal of the American Society for Information Science, 45,* 20–30.

Morrow, R. A., & Torres, C. A. (2002). *Reading Freire and Habermas: Critical pedagogy and transformative social change.* New York: Teachers College Press.

Muller, M. J., & Czerwinski, M. (1999). Organizing usability work to fit the full product range. *Communications of the ACM, 42*(5), 87–90.

Nelson, H. (2005). *Interview*. Retrieved June 2, 2006, from http://www.distance-educator.com/dnews/modules.php?op=modload&name=News&file=article&sid=13424

Nichols, R. G., & Allen-Brown, V. (1996). Critical theory and educational technology. In D. H. Jonassen (Ed.), *Handbook of research for educational communications and technology* (pp. 226–252). New York: Macmillan Library Reference U.S.A; Simon & Shuster Macmillan

Norman, D. A. (1983, Dec.). *Design principles for human–computer interfaces*. Paper presented at the Conference on Human Factors and Computing Systems, Boston, MA.

Norman, D. A. (1989). The electronic library: How will the user cope. *Bulletin of the American Society of Information Science, 15*(5), 8–9.

Norman, D., & Draper, S. W. (1986). *User centered system design: New Perspectives on human–computer interaction*. Hillsdale, NJ: Lawrence Erlbaum Associates, Inc.

Norrbom, M. (2001, April). Employment law in Denmark. *International Financial Law Review, 23*–26. Supp/2

Norum, K. E. (2000). *Appreciative instructional design (AiD): A new model for a new millennium*. Unpublished manuscript.

Noyes, J. M., & Baber, C. (1999). *User-centered design of systems*. New York: Springer.

Olson, L. (1990). "Jury still out" on re: Learning's grassroots reform experiments. *Education Week, 10*(5), 44.

Otten, M. (1991). Changing the workplace to fit human needs: The Norwegian work environment act. *Economic and Industrial Democracy, 12*(4), pp. 487–500.

Park, P., Brydon-Miller, M., Hall, B., & Jackson, T. (1993). *Voices of change: Participatory research in the United States and Canada*. Westport, CT: Bergin & Garvey.

Patton, M. Q. (1990). *Qualitative evaluation and research methods* (2nd ed.). Newbury Park, CA: Sage.

Payette, S. D., & Reiger, O. Y. (1998). Supporting scholarly inquiry: Incorporating users in the design of the digital library. *Journal of Academic Librarianship, 24*, 121–129.

Pena, D. C. (2000). Sharing power? An experience of Mexican American parents serving on a campus advisory council. *The School Community Journal, 10*, 61–84.

Petroski, H. (1994). *Design paradigms: Case histories of error and judgement in engineering*. Cambridge, England: Cambridge University Press.

Porter, M.E. (1985) *Competitive advantage: creating and sustaining superior performance*. NY: Free Press

Postman, N. (1992). *Technopoly:_ The surrender of culture to technology*. New York: Knopf.

Prigogine, I., & Stengers, I. (1984). *Order out of chaos*. New York: Bantam.

Purpel, D. E. (1989). *The moral and spiritual crisis in education: A curriculum for justice and compassion in education*. South Hadley, MA: Bergin & Garvey.

Quade, E. S. (1963). Military systems analysis. In S. L. Optner (Ed.), *Systems analysis* (pp. 121–140). Harmondsworth, England: Penguin.

Rapoport, A. (1986). *General systems theory: Essential concepts and applications*. Cambridge, MA: Abacus Press.

Reason, P. (1994). Three approaches to participatory inquiry. In N. K. Denzin & Y. S. Lincoln (Eds.), *Handbook of qualitative research* (pp. 324–339). Thousand Oaks, CA: Sage.

Reeves, T. C. (2005). Design-based research in educational technology; progress made, challenges remain. *Educational Technology, 45*(1), 48–52.

Reiber, L. (1998) The proper way to become an instructional technologist 1998 Peter Dean Lecture for the division of learning and performance environments. Presented at the Annual Conference for Educational Communications and Technology. St. Louis, February.

Reigeluth, C. M. (1993). Principles of educational systems design. *International Journal of Educational Research, 19,* 117–131.

Reigeluth, C. M. (1996). A new paradigm of ISD? *Educational Technology, 36*(3), 13–20.

Reigeluth, C. M., & Garfinkle, R. J. (1992). Envisioning a new system of education. *Educational Technology, 32*(11), 17–23.

Riley, M. S. (1986). User understanding. In D. A. Norman & S. W. Draper (Eds.), *User centered system design* (pp. 157–170). Hillsdale, NJ: Lawrence Erlbaum Associates, Inc.

Rockman, I. F. (1980). The potential of on-line circulation systems as public catalogs: An introduction. RQ: Reference Quarterly, 20, 39–58.

Rogers, E. M. (1995). *Diffusion of innovations.* New York: Free Press.

Roma, C. (1990). *Formative evaluation research on an instructional theory for understanding.* Unpublished doctoral dissertation, Indiana University Graduate School, Bloomington, IN.

Romanish, B. (1993). Teacher empowerment as the focus of school restructuring. *The School Community Journal, 3,* 47–60.

Ross, S.M. & Morrison, G.R. (1993) How to get research articles published in professional journals. Tech Trends, 38(2), 29–33.

Rothwell, B. (2005). *Beyond training and development: The groundbreaking classic on human performance enhancement.* New York: American Management Association.

Rowland, G. (1993). Making change our friend: The design perspective. *Educational Technology, 33*(7), 29–31.

Sachs, S.G. (1984). Citation Patterns in instructional development literature. Journal of Instructional Development, 7(2), 8–13.

Salisbury-Glennon Jill. D., Gorrell Jeffrey, Sanders Steve, Boyd Pamela, Kamen-Michael (April 1999) Self regulated learning strategies used by the learners in a learner-centered school. Paper presented at the annual meeting of American Educational Research Association, Montreal, Quebec, Canada.

Salvo, M. J. (2001). Ethics of engagement: User-centered design and rhetorical methodology. *Technical Communication Quarterly, 10,* 273–290.

Sarason, S. B. (1995). *Parental involvement and the political principle: Why the existing governance structure of schools should be abolished.* San Francisco: Jossey-Bass.

Sarason, S.B. & Lorentz, E.M. (1998). Crossing boundaries: Collaboration, Co-ordination and the Redefinition of Resources. Jossey-Bass, San Francisco.

Schein, E. H. (1999). Cultural change. In D. G. Langdon, K. S. Whiteside, & M. M. McKenna (Eds.), *Intervention resource guide: 50 performance improvement tools* (pp. 125–130). San Francisco: Jossey-Bass.

Schoemaker, P.J.H. (1995) Scenario Planning: A tool for strategic thinking: Sloan Management Review Winter, pp. 25–40

Schon, D. A. (1983). *The reflective practitioner: How professionals think in action.* New York: Basic Books.

Schuler, D., & Namioka, A. (1993). *Participatory design: Principles and practices.* Hillsdale, NJ: Lawrence Erlbaum Associates, Inc.

Schulze, A. N. (2001). User-centered design for information professionals. *Journal of Education for Library and Information Science, 42,* 116–122.

Seels, B., & Glasgow, Z. (1990). *Exercises in instructional design.* Columbus, OH: Merrill.

Senge, P. (1990). *The fifth discipline: The art & practice of the learning organization.* New York: Doubleday.

Sharp, D. L. M., Bransford, J. D., Goldman, S. R., Risko, V. J., Kinzer, C. K., & Vye, N. J. (1995). Dynamic visual support for story comprehension and mental model building by young, at-risk children. *Educational Technology Research and Development, 43*(4), 25–42.

Skinner, B. F. (1961). *The analysis of behavior: A program for self instruction*. New York: McGraw Hill.

Skolimowski, H. (1985, April). *The co-creative mind as partner of the creative evolution*. Paper presented at the 1st international conference on Mind–Matter Interaction, Universidad Estadual de Campinas, Brazil.

Soloway, E., & Pryor, A. (1996). The next generation in human–computer interaction. *Communication of the ACM, 39*(4), 16–18.

Stepien, W., & Gallagher, S. (1993). Problem-based learning: As authentic as it gets. *Educational Leadership, 50*(7), 25–28.

Stevenson, K. R., & Pellicer, L. O. (1992). School-based management in South Carolina: Balancing state-directed reform with local decision making. In J. J. Lane & E. G. Epps (Eds.), *Restructuring the schools: Problems and prospects* (pp. 123–139). Berkeley, CA: McCutchan.

Stolovitch, H. D., & Keeps, E. (1992). *Handbook of human performance technology: A comprehensive guide for analyzing and solving performance problems in organizations*. San Francisco: Jossey-Bass.

Stringer, E. T. (1996). *Action research: A handbook for practitioners*. Thousand Oaks, CA: Sage.

Sugar, W. A. (1995). Impact of user-centered design methodology on the design of information systems. *Proceedings of the ASIS Annual Meeting, 32*, 211–214.

Sugar, W. A. (2000, Oct.). *Technology bill of rights for educators*. Paper Presented at the 2000 annual convention of the Association of Educational Communications and Technology, Denver, CO.

Sugar, W. A. (2001). What is so good about user-centered design? Documenting the effect of usability sessions on novice software designers. *Journal of Research on Computing in Education, 33*, 235–250.

Sugar, W. A., & Boling, E. (1995, Feb). *User-centered innovation: A model for "early usability testing."* Paper presented at the annual national convention of the Association for Educational Communications and Technology, Anaheim, CA.

The Homeowner Tax Relief Act (2004). The Commonwealth of Pennsylvania.

The Homeowner Tax Relief Act, (July, 2004) Pennsylvania.

Thoresen, J. D. (1984). Using simulated job samples for skills evaluation: Learning from the Assessment Center method. *Performance and Instruction, 23*, 16–18.

Torbert, W. R. (1981). Why educational research has been so uneducational: The case for a new model of social science based on collaborative inquiry. In P. Reason & J. Rowan (Eds.), *Human inquiry: A sourcebook of new paradigm research* (pp. 141–152). Chichester, England: Wiley.

Tripp, S., & Bichelmeyer, B. (1990). Rapid prototyping: An alternative instructional design strategy. *Educational Technology Research and Development, 38*(1), 31–44.

Urbanski, A. (1995) Learner-centered schools: A vision for the future. *Educational Policy, 9*, 281–292.

Valente, T. W., & Davis, R. L. (1999). Accelerating the diffusion of innovations using opinion leaders. *Annuals of the American Academy of the Political and Social Sciences, 566*, 55–67.

Van der Heijden, K. (1997). *Scenarios, strategies and the strategy process*. Breukelen: Nijenrode University Press.

Vidich, A. J., & Lyman, S. M. (1994). Qualitative methods: Their history in sociology and anthropology. In N. Denzin & Y. Lincoln (Eds.), *Handbook of qualitative research* (pp. 23–59). Thousand Oaks, CA: Sage.

Volk, T. (1995). *Metapatterns: Across space, time, and mind*. New York: Columbia University Press.

REFERENCES

von Bertalanffy, L. (1968). *General systems theory.* New York: Braziller.
von Glasersfeld, E. (1995). *Radical constructivism: A way of knowing and learning.* Washington, DC: Falmer.
Wagner, E. D., & McCombs, B. L. (1995). Learner centered psychological principles in practice: Designs for distance education. *Educational Technology 35*(2), 32–35.
Warfield, J. N. (1994). *A science of generic design: Managing complexity through systems design.* Ames: Iowa State University Press.
Weinberger, E. McCombs Barbara, L. (April, 2001) The impact of learner-centered practices on the academic and non-academic outcomes of upper elementary and middle schools students. Paper presented at the annual meeting of the American Educational Research Association, Seattle.
Wiggins, G. P. (1998). *Understanding by design.* Alexandria, VA: Association for Supervision and Curriculum Development.
Wilson, L. A. (1995). Building a user-centered library. *RQ,: Reference Quarterly 34,* 297–301.
Wilson, L. A., & Arp, L. (1995). Library literacy: Building the user-centered library. *RQ: Reference Quarterly, 34,* 297–302.
Winner, L. (1986). *The whale and the reactor: A search for limits in an age of high technology.* Chicago: University of Chicago Press.
Wydra, F. T., & McKenna, M. M. (1999). Learner-controlled instruction. In D. G. Langdon, K. S. Whiteside, & M. M. McKenna (Eds.), *Intervention resource guide: 50 performance improvement tools* (pp. 204–210). San Francisco: Jossey-Bass.
Yaffe, E. (1994). Not just cupcakes anymore: A study of community involvement. *Kappan,* pp. 697–705.
Yapa, L. (1996a). Innovation diffusion and paradigms of development. In C. Earle, K. Mathewson, & M. Kenzer (Eds.), *Concepts in human geography* (pp. 231–270). Lanham, MD: Rowman & Littlefield.
Yapa, L. (1996b). What causes poverty? A postmodern view. *Annals of the Association of American Geographers, 86,* 707–728.
Zemke, R. (1989). The systems approach: A nice theory but … In C. Lee (Ed.), *Performance technology* (pp. 19–22). Lakewood Books.

Author Index

A
Abels, E.G., 9
Adler, P.A., 25, 30
Adler, P., 25, 30
Allen-Brown, V., 52
AML, 21
Apple, M. W., 52
Argyris, C., 35
Arp, L., 17
Atkinson, P., 29

B
Baber, C., 15
Banathy, B.H., 4, 5, 16, 22, 66, 87, 91, 92, 93, 108
Bansler, J., 20
Barabasi, A.L., 94
Barrows, H. S., 9, 10
Bateson, G., 88
Bauch, P.A., 4
Bell, A., 100, 102
Benavente, A., 18
Berube, M., 22
Bhola, H. S., 3
Bichelmeyer, B., 33
Blomberg, J., 30, 31
Bohm, D., 26, 98
Bodker, S., 20, 32, 33, 83
Boulding, K. E., 86, 89
Breman, J., 58
Briggs, L. J., 86
Bromley, H., 52
Brown, J. S., 2, 16
Brydon-Miller, M., 25, 35
Burns, J., 2, 16

C
Capra, F., 86
Carey, L., 2, 53, 86, 87

Carleton, J. R., 2
Carmen, R., 7
Carr, A.A., 4, 21, 25, 63, 72, 73, 85, 111, 114
Carr, I.C., 18
Carr-Chellman, A.A., 2, 5, 6, 58, 65, 71
Carr-Chellman, D.J., 71
Carroll, J.M., 39
Carspecken, P.F., 28
Checkland, P., 22, 23, 86, 90, 91
Chomsky, N., 94, 95
Churchman, C. W., 23, 86
Cochran, M., 4
Cognition and Technology Group at Vanderbilt, 10
Comer, J.P., 4
Comstock, D. E., 35
Cooper, B.S., 22
Coupland, D., 78
Cuyar, C., 58

D
Damarin, S.K., 18
Daniels, D.H., 16
Daresh, J.C., 21, 26
Davidson, M., 90
Davies, D., 22
Davydov, V. V., 9
Dean, C., 4
DeCandido, G.A., 17
Delgado-Gaitan, C., 4
Demming, W. E., 86
Design-Based Research, 38
Dick, W., 2, 53, 86, 87
Dillon, A., 9
Draper, S.W., 9, 17
Dreikorn, M.J., 104
Duffy, T. M., 9, 10, 16
Duguid, P., 2, 16

AUTHOR INDEX

E
Epstein, J.L., 22
Ertmer, P., 53

F
Fantini, M., 22
Fidel, R., 17
Filler, M., 70
Flood, R.L., 23, 51, 52
Floyd, J.D., 60
Fontana, A., 25, 30
Forbes, R., 100, 102
Ford, J.K., 100, 102, 103
Fox, R., 35
Frey, J. H., 25, 30

G
Gagne, R. M., 86
Gallagher, S., 9
Gardner, J. W., 87
Garfinkle, R.J., 23
Garrison, J., 9
Gayeski, D., 2
Gerhardt, H.P., 18
Giacomi, J., 30, 31
Gilbert, T. F., 2
Giroux, H.A., 52
Gittell, M., 22
Glasersfeld, Ernst von, 10
Glasgow, Z., 53, 86
Goldring, E.B., 4
Gronbaek, K., 20, 32, 33, 83

H
Habermas, J., 52
Hall, A. D., 25, 26, 35, 86
Hall, B., 25, 26, 35
Hammersley, M., 29
Hannafin, M. J., 2, 9
Harman, W. W., 86
Haynes, N.M., 4
Heidegger, M., 25
Henderson, L., 2
Hirschkop, K., 94, 95
Hitch, C. J., 86
Hoadley, C.M., 2, 26, 38
Holloway, L. D., 9
Horn, R. A., 72
Horton, B. D., 35
Hlynka, D., 50
Hurley, D.E., 9
Hutchins, C. L., 86

I
IDT Futures Group, 111
ISSS (International Society of Systems Sciences), 72, 88, 93

J
Jackson, T., 16, 25, 26, 35
Jantsch, E., 23
Jenkins, G. M., 86
Jenlink, P. M., 3, 4, 5, 22, 73, 74, 114
Jonassen, D.H., 9, 10, 16, 107
Jones, J. C., 87, 111

K
Kalkman, D.L., 1
Keeps, E., 2
Kelley, R., 78, 79
Kelly, K., 5, 21
Klein, J. D., 111
Kyng, M., 20, 32, 33, 83

L
Land, S. M., 2, 9
Larsen, T.J., 7
Laurillard, D., 2
LeCompte, M. D., 30
Lineberry, C., 2
Lippett, L., 39, 70
Lowyck, J., 16
Lyman, S. M. 28

M
MacDonald, B., 69
Magat, R., 22
Maher, F., 2
McCandless, P., 17
McChesney, R.W., 94, 95
McCombs, B.L., 16
McGuire, E., 7
McKenna, M.M., 103
McLaren, P., 52
Melo, A., 18
Merrifield, J., 15
Midgley, C., 4
Miller, J.G., 22
Ming-fen, L., 22
Morris, R., 17
Morrison, G., 111
Morrow, R.A., 18
Mosher, A., 30, 31

N
Namioka, A., 6, 17, 20, 26
Nelson, H., 107

AUTHOR INDEX

Nichols, R.G., 52
Norman, D.A., 9, 17
Norrbom, M., 21
Norum, K.E., 53
Noyes, J.M., 15

O
Olson, L., 15
Otten, M., 21

P
Park, P., 25, 26, 35
Patton, M. Q., 35
Payette, S.D., 18
Pellicer, L.O., 21
Pena, D.C., 4
Petroski, H., 87
Preissle, J., 30
Prigogine, I., 23
Postman, N., 25
Purpel, D.E., 96
Putnam, R., 35

Q
Quade, E. S., 86
Quait, M., 16

R
Rapoport, A., 22
Reason, P., 35
Reeves, T. C., 36
Reigeluth, C. M., 3, 4, 5, 22, 23, 59, 65, 111, 114
Reiger, O.Y., 18
Richey, R., 111
Riley, M.S., 17
Rockman, I.F., 17
Rogers, E. M., 37, 46
Roma, C., 36
Romanish, B., 4
Romm, N., 23, 51, 52
Ross, S., 111
Rosson, M.B., 39
Rothwell, B., 101
Rowland, G., 2, 87

S
Sachs, S., 111
Salvo, M., 15

Sarason, S. B., 22, 26, 95
Schein, E.H., 101, 103
Schon, D. A., 35
Schuler, D., 6, 17, 20, 26
Schulze, A.N., 15
Seels, B., 53, 86, 111
Senge, P., 16, 100
Silber, K. H., 111
Skolimowski, H., 88
Smith, M.C., 35
Stengers, I., 23
Stepien, W., 9
Stevenson, K.R., 21
Stolovitch, H. D., 2
Stringer, E. T., 35
Sugar, W.A., 9, 15, 53
Swenton-Wall, P., 30, 31

T
Tetreault, M., 2
Thoresen, Joseph D., 60
Torbert, W. R., 35
Torres, C.A., 18
Tripp, S., 33

V
Vaner Heijden, K., 41
Vidich, A. J., 28
Volk, T., 88
Von Bertalanffy, L., 22, 86

W
Wager, W. W., 86
Wagner, E.D., 16
Warfield, J.N., 87
Weinberger, E., 16
Whistler, J.S., 16
Wiggins, G.P., 52
Willson, L.A., 17
Winner, L., 25
Wood, S., 4
Wydra, F.T., 103

Y
Yaffe, E., 49
Yapa, L., 7

Z
Zemke, R., 105

Subject Index

A
Action research, *see also* Action research based UD, 26–28, 33–36, 43, 52, 61, 64
Action research based UD *see also* Action research, 25, 27, 33–36, 43
ADDIE (model), *see also* Instructional design, 112
Adoption, *see also* Implementation, 3, 7–8, 12, 14–15, 21, 26, 40, 46, 58, 66, 102, 105, 110, 121, 123
Agoras, 72
Anthropology, 30
Apathy, 24, 49–52, 57

B
Behaviorism, 7
Bill Gates, 78, 96
Bill O'Reilly, 73
Buy-in, 26

C
Capitalism, 95
Cartesian model, 97
Change, *see also* Innovation, 13, 24, 27, 35, 38, 48. 58, 62–64, 69, 75, 77, 80, 85–98, 102, 104, 107, 123
Chaos, *see also* Dynamical systems, 50, 100, 109
Charter schools, 74–75
Christmas Tree change, 64
Classism, 95
CNN, 86
Coaching, 101–102
Community, 35
Conflict, 57–76, 104, 106, 121, 123
Confucianism, 78
Consensus, 71
Constructivism, 2, 9–10, 46

Context 2, 21, 28, 41, 54, 59 77, 85, 87, 95
Conversation, 26
 design, 65, 73–76
 groups, 64–70
Cooperative design (as UD tool), *see also* Cooperative prototyping, 25, 26, 31–33, 43
Cooperative prototyping *see also* Cooperative design, 32–33, 62
Cost-benefit, *see also* ROI, 61, 77, 79, 83, 104, 109
Critical
 perspective, 18–19, 23, 38–39
 systems theory, 51
 theory, 52
Culture, 26

D
Decision-making, 53 60, 62
Democracy, 20, 72, 88. 94–96
Democratic, 12, 50–52, 72, 94–96
Design, 2, 25, 27 33, 35–36, 39, 47–48, 50–51, 65, 87, 107–
Design based research (DBR), 26–27, 36–39, 44
Dialogue, 72–73
Diffusion, 3
Diversity, 46–47, 66–67 71, 74–75
Dynamical systems, *see also* Chaos, 22, 26

E
Educational Systems Design (ESD), 49, 51, 59, 61, 91–93, 108
Emancipatory design, 18–19, 23, 38–39
Embeddedness, 86, 88, 100
Epistemology, 92, 106
Empowerment, 18–19, 22, 26, 27, 50, 58, 63, 84, 101, 109–110, 121, 123

SUBJECT INDEX

Enabling systems, 59
Ethnography (as UD tool), 25, 27–31, 36, 43
Evolutionary prototyping, 60
Expertism, 6, 22, 27, 35, 41, 44, 47–49, 52–54, 63, 68, 85

F
Facilitation, 35, 45–57, 63, 121
Failures (predictable), 95, 98
Field work, 28–31
Formative research, 36
Fractals, 91

G
Generalizability, *see also* Positivism 28, 36, 69, 85
Global perspective, *see also* Holism, 86, 88
Grassroots movements, 21, 77

H
Heteropholy, *see also* Homophily, 63
Holism, 22, 26–27, 85, 88
Homophily, *see also* Heteropholy, 63
Human computer interface (HCI), 6, 16–17, 20, 23, 39, 64, 108

I
Implementation, *see also* Adoption, 3, 10, 13, 21–23, 33, 40, 58–60, 66, 79, 82, 97, 103, 111, 115, 118, 121, 123
Indigenous knowledge, 22, 27, 35, 62, 63, 68, 80
Information age, 3, 45
Innovation, *see also* Change, 3, 14, 21, 63
Instructional design, 1, 2, 4, 6–8, 23, 27–28, 45–48, 52–53, 79, 86–87, 97, 99, 103, 106–123
  models, 52–53, 106, 112
  field identity, 111
Interconnectedness, 22, 86, 88, 100
Interdependency, 22 86, 100
Internet, 5–6, 94–95, 98
Iterative (nature), 110

K
Koi (fish), 78

L
Leader
  autocratic, 79
  buy in, 55, 59 60, 62–63, 65, 67–68
  ship, 24, 41, 55, 58, 63, 67, 77–98, 103, 105, 106
  transformational, 79

Learner-centered, *see also* User-centered, 10 16–17, 49
Lerner controlled instruction, 103
Learning, 10, 53, 73, 80 106
Learning Sciences, 2, 36
Learning theories, 108
Liberatory systems theory, *see also* Emancipatory design, 22
Library, 17
Linear, *see also* Systematic, 85

M
Mandated change, 14, 51, 69, 80, 96
Media usage, 17
Microsoft, 78, 95–96
Moral purpose, 72, 123

N
Needs assessment, 4
No Child Left Behind (NCLB), 10, 14

O
Ontology, 92
Open Space Technology, 70
Organizational change, *see also* Performance technology 14, 21, 86, 99–204
Ownership, 7–8, 32, 46, 102

P
PA Act 72, 50–51
Participatory design, 6, 20, 26
Participatory rural appraisal (PRA), 27–28
Performance Technology, 2, 99–105
Piecemeal change, 86, 89
Political nature, 32–33
Positivism, *see also* generalizability, 28, 36
Power, 6–10, 15, 19, 24, 26–27, 46–51, 53, 55, 57–58, 61–62, 67–68, 80–81, 83, 99
Preferred futuring, 39, 70
Pyramid teams, 60, 65

R
Racism, 95
Resistance, 11, 26,101, 104
Rapid prototyping, 3
Return on Investment (ROI), 61, 77, 79, 83, 104
Ripple effects (of changes), 86
Rush Limbaugh, 73

S

Scandinavian traditions, 6 19–21, 23, 32, 77, 110
Scenario based user design, 26–27, 39–41
Scenario planning, 26–26, 39–41, 44
School change, 21, 48, 64–70
Scientific reductionism, 91
Self-correction, 93
Self-reflection, 35–36
Site-based decision making (SBDM), 26
Social Security System crisis, 70–71
Social change, 68
Specialization, 90
Spiral development method, 60
Spiritual age, 45–46
Stakeholder
  defined, 4
  involvement, 3–5, 21, 46–48
  participation, 19, 21–22, 60, 62, 65, 66, 68, 100–101
Systematic, 85–86, 89, 92–93, 105, 110, 123
Systems change, 1, 26, 35, 47, 69, 74, 85–98, 105
System-of-interest, 86–87
Systems theory, 1, 19, 22–23, 87–94, 99, 100
  hard, 22, 52, 86, 110
  living, 22
  soft, 86, 90–91
  social, 21–23, 52
Subsystem, 86–87
Suprasystem, 86–87

T

Team membership, 66–67, 80–81
Total Quality Management (TQM), 26

Town meetings, 72
Turnover (in team membership), 66

U

User
  commitment, 70
  defined, 15
  participation, 15, 21, 23
  responsibility, 88
User-centered Design, *see also* Learner-centered 8–10, 15–17, 20, 47
User Design
  appropriateness, 10–12, 26–28, 77, 104–105
  advantages/disadvantages, 12, 61, 79–80, 83 104
  cases, 29–30, 31–32, 34–40, 58–71, 116–121
  definition, 1, 2, 5–10
  ethics of, 45–46
  field identity, 111–112
  impact, 6–8
  obstacles to, 23–24, 49–54, 57–76, 79, 82
  origins, 19–23
  process, 54, 70, 77, 112–116
  tools for, 25–44, 55, 63, 70

V

Value (non-neutrality of), 25

W

Waterfall development method, 60
World Trade Center (as user design), 70